BIOTECHNOLOGY AND THE HUMAN GOOD

BIOTECHNOLOGY AND THE HUMAN GOOD

C. Ben Mitchell, Edmund D. Pellegrino,
Jean Bethke Elshtain, John F. Kilner,
and Scott B. Rae

Georgetown University Press / Washington, D.C.

As of January 1, 2007, 13-digit ISBN numbers have replaced the 10-digit system.

13-digit	10-digit
Paperback: 978-1-58901-138-0	Paperback: 1-58901-138-4

Georgetown University Press, Washington, D.C. www.press.georgetown.edu

Biotechnology and the human good / C. Ben Mitchell . . . [et al.].
p. ; cm.
Includes bibliographical references and index.
ISBN 1-58901-138-4 (alk. paper)
1. Biotechnology—Moral and ethical aspects. 2. Medical ethics.
3. Values. I. Mitchell, C. Ben.
[DNLM: 1. Biotechnology—ethics. 2. Christianity. 3. Genetic Engineering—ethics. TP 248.23 B615 2006]
TP248.23.B566 2006
174′.96606—dc22 2006021475

∞ This book is printed on acid-free paper meeting the requirements of the American National Standard for Permanence in Paper for Printed Library Materials.

14 13 12 11 10 09 08 9 8 7 6 5 4 3 2

Printed in the United States of America

Contents

Preface

LISTENING to Kevin Warwick, a professor of cybernetics at the University of Reading in England, enthuse about his research is like listening to a prepubescent schoolboy describing his trip to Disney World. Warwick claims to be the world's first cyborg: part human, part machine. On Monday, August 24, 1998, he had a silicon chip transponder surgically implanted in his forearm. Once fitted with this new implant, he returned to his laboratory, where the doors opened automatically, lights turned on as he walked into rooms, and his computer greeted him every morning. As thrilling as this was, it was only the beginning.

In March 2002, Warwick embarked on Project Cyborg 2.0. This time, surgeons at Oxford's Radcliffe Infirmary implanted a 100-microelectrode array directly into the median nerve fibers of his left arm. This new device allowed the professor's nervous system, including his brain, to be connected directly to a computer. As a result, Warwick was able to control a robotic arm in his lab, drive an electric wheelchair with minimal hand movement, and, through a secret Internet connection, control an articulated robotic arm on another continent. He was both able to send signals across the ocean and receive them directly into his nervous system.

Warwick's wife, Irena, volunteered to have a similar implant placed in her wrist, allowing husband and wife to "communicate" directly through the computer, thereby becoming the world's first cyborg couple.

Warwick's excitement about his experiments is almost overwhelming, and for good reason. This new technology may one day be used to treat many types of neuromuscular disorders. A robotic prosthesis may restore arms, legs, or other appendages lost through injury. Eventually, says Warwick, the technology may allow us to communicate our emotional states directly to another person. But the same technology may also be used to create armies of efficient cyborg killers. With the computational power of a laptop computer and a rather basic knowledge of microbiology, one can now manipulate living organisms in one's own basement lab, creating who knows what? Technology, including biotechnology, may be used for evil ends as easily as for good purposes, and this worries Warwick. But what is one to do? Technology marches on.

If we understand technology to include any work-producing extension of the body of an individual, then the first person to use a stick to make a hole in which to plant a seed was a technologist. In fact, tool making was one of the first human technological advances. Human beings are by nature technologists. Therefore, biotechnology is a fundamentally human endeavor.

Nevertheless, when the media announce a new biotechnological development—such as the possibility of human cloning or the creation of animal–human hybrids—there is a collective gasp. How do we account for what some might describe as our schizophrenic reactions to biotechnology? How do we test our intuitions about emerging biotechnologies?

The University of Montana philosopher of technology Albert Borgmann has wryly observed that reactions to emerging biotechnologies, including cybernetics and artificial intelligence, "are as divided as they are to carnival rides—they produce exhilaration in some people and vertigo in others" ("On the Blessings of Calamity and the Burdens of Good Fortune," *Hedgehog Review* 4 [Fall 2002]: 7–24).

Techno-exhilaration and techno-vertigo are intuitional responses to these new technologies. Some technologies call on our adrenal glands to work overtime because of the breathtaking nature of new power within our grasp. From biplanes, to space flight, to moon walks, the human heart races with anticipation at the next great achievement.

From blood transfusions, to the discovery of DNA, to gene therapy, the passion for knowledge pushes us into new frontiers. At the same time, it seems, every power of mastery over nature also brings with it greater power for mastery over other human beings, and every inch closer to technological utopia seems to be another step toward technological oblivion.

This book is the effort of a multidisciplinary group of physicians, scientists, philosophers, ethicists, theologians, and a lawyer to grapple with these questions and to offer a way of thinking about technology—especially biotechnology—that we hope will make sense of some of our intuitions. However, a warning is in order. History has shown that some of our intuitions about biotechnology are wrong and should be questioned. For instance, in the seventeenth century, blood transfusions were outlawed in France and England. Now more than 39 million units of blood and blood products are transfused every year in the United States alone. The intuition that it was wrong to transfer blood—the elixir of life, as it is sometimes called—has been revised over time. (For an intriguing account of this history, see Pete Moore, *Blood and Justice: The 17th Century Parisian Doctor Who Made Blood Transfusion History* [San Francisco: Jossey-Bass, 2002].) So part of the purpose of this book will be to test our intuitions about biotechnology.

Chapter 1 surveys the rapidly expanding arena of human biotechnology. Technologies such as genetic manipulation, cybernetics, robotics, and nanotechnology not only offer great hope for therapeutic interventions but also portend potentially devastating challenges to our understanding of what it means to be human and, in some instances, to our humanity itself. Genetic enhancements, some argue, may lead human beings to a technologically achieved immortality. But at what cost to our humanity?

Chapter 2 critiques several narrative philosophies that offer arguments for or against technological expansion. Though by no means the only narrative, what the American studies professor David E. Nye calls the "second-creation narrative" seems to be a dominant theme of the Western story of technological achievement. Without modification, this narrative seems to be an insufficient ground for our biotechnological agenda. A narrative of morally responsible stewardship, however, offers both an impetus for biotechnology and realism about the potential for misuses of biotechnology. In our view, biotechnology is a qualified good and should be pursued with vigor, but not without caution. We are not determinists when it comes to biotechnology. We do not believe that because we *can* do something we *ought* to do it; or that once we *can* do something, it is inevitable that we *will* do it. On the contrary, as stewards, we believe biotechnology should be used to relieve human suffering and to protect human dignity, without relieving humans of their very humanity.

Chapter 3 explores several competing worldviews that inform our attitudes toward biotechnology. Philosophical naturalism has almost imperialistic status in the sciences in general and in biotechnology in particular. We argue that philosophical naturalism is limited and reductionist. Alternatively, though environmentalist biocentrism, or deep ecology, is more protective of living organisms, the movement leads to undervaluing the human species vis-à-vis other species. We maintain that Judeo-Christian theism offers a more satisfying way to frame the goals of biotechnology because it offers a view of human dignity, combined with a purposive history, that warrants therapeutic applications of biotechnology without either sanctioning the wholesale modification of the human species or overprotecting the environment to the detriment of the human species.

In chapter 4, we elucidate a view of human dignity that ought to underwrite the biotechnological enterprise. The notion of human dignity emerges from a lengthy, and sometimes painful, history in the West and has significantly shaped our understanding of human rights, including the protection of human subjects in research. Assaults against human dignity have led to some of the darkest days in history, including American chattel slavery, the Holocaust in Nazi Germany, and violations of human rights in scientific experimentation. Only a robust understanding and protection of human dignity can prevent us from repeating the horrors of the past. When properly understood, human dignity is—or can become—a shared value for informing law and policy.

Does the idea of human dignity entail complete control over one's future destiny, including, if one freely chooses, altering or even jettisoning one's body? Chapter 5 examines the contemporary penchant for autonomous control that seems to be driving many of those who are uncritical proponents of biotechnology. A renewed appreciation for human embodiment and community would resist the neo-Gnostic tendencies of twenty-first century biotechnology.

Chapter 6 provides a historical and conceptual framework for thinking about the nature of morally responsible stewardship in relation to the goals of medicine. Are enhancement technologies consistent with the teleology of medicine? We argue that they are not. Moreover, we maintain that some biotechnologies fatally compromise the physician–patient relationship, turning patients into consumers and physicians into mere contractors. Medicine ought to resist being co-opted by a narcissistic, consumerist culture.

In chapter 7, we offer a series of profound questions that must be answered if technology is to serve human needs and goals. In the last section of the chapter, we bring a philosophical and theological framework to bear on biotechnology. We believe that it is not only appropriate but also necessary to question biotechnology—not to stop it, but to keep it honest. We also believe that by asking the hard questions about biotechnology, we can move together toward a truly human future in which therapy, healing, and health can be preserved without forfeiting medicine to the whims of a dysfunctional utopianism.

Acknowledgments

READERS will notice that we do not identify an author for each chapter of this book. This directly reflects our methodology. After the book's authors and external reviewers were chosen, the team met for a two-day retreat. Each of us presented our initial thoughts about the goals and content of the project. After extensive critiques, we agreed on a trajectory for the book. Next, each author presented a draft manuscript on one of the topics for the book. The drafts were reviewed by each author; by the project director, C. Christopher Hook, M.D.; and by several external reviewers, including Paige Comstock Cunningham, J.D., and Gilbert C. Meilaender, Ph.D. In addition, two reviewers outside North America commented on the essays—from Australia, Graham Cole, Ph.D., and from Ireland, Stephen Williams, Ph.D.—providing welcome and helpful international perspectives.

Additional feedback was received when second drafts of the material were presented to a larger audience during the "Remaking Humanity?" conference sponsored by the Center for Bioethics and Human Dignity in Bannockburn, Illinois. The ongoing conversation and collaboration continued as the team met twice to review the external feedback and self-critiques. The final review was completed early in 2006. We would especially like to thank the outside commentators for helping us focus our thinking. Without their extraordinarily helpful comments, the book would not have come together as it has.

In sum, this book is the product of substantial longitudinal collaboration and conversation, with each person making significant contributions to each chapter of the whole.

We are very grateful for the financial support this project has received from the William H. Donner Foundation, New York, and from Fieldstead & Company, Irvine, California, and for administrative support from the Center for Bioethics and Human Dignity.

Our deepest thanks go to the project editor, Louise Kaegi, M.A., who gave the book one voice. Louise writes on health care, ethics, education, and cultural politics, was formerly executive editor of the Joint Commission on Accreditation of Healthcare Organizations' *Joint Commission Benchmark* newsletter, and has written articles on health care for other newsletters, such as *Minority Nurse*. We would also like to thank Michael Sleasman, a doctoral candidate and friend at Trinity Evangelical Divinity School, in Deerfield, Illinois, for his bibliographic assistance with a part of the project.

This book belongs largely in the genre of bioethics. Bioethics may be the most salient contemporary example of interdisciplinary research. If there were ever a time when one scholar, working solely in her office, could become an expert on all the topics under her discipline, that day has passed. With the explosive development of biotechnology, the burgeoning growth of specialties and subspecialties, and the large output of literature, collaborative research will be increasingly necessary. We applaud Chris Hook's vision for the type of deep collaboration represented in this volume. Likewise, we are grateful to Georgetown University Press and its director, Richard Brown, for recognizing the importance of this work. We hope that other presses will understand how important collaborative research is in this area and will support colleagues in their joint labors in bioethics.

CHAPTER ONE

The Rapidly Changing World of Biotechnology

> We reduce things to mere Nature *in order that* we may "conquer" them. We are always conquering Nature, *because* "Nature" is the name for what we have, to some extent, conquered. The price of conquest is to treat a thing as mere Nature. . . . As long as this process stops short of the final stage, we may well hold that the gain outweighs the loss. But as soon as we take the final step of reducing our own species to the level of mere Nature, the whole process is stultified, for this time the being who stood to gain and the being who has been sacrificed are one and the same.
>
> —C. S. Lewis, *The Abolition of Man*

TECHNOLOGY is an integral part of human life, from the simplest tools made from plants and stones to our digital computers, cardiac pacemakers, pharmaceuticals, and the ubiquitous media of communication and transportation. Human beings are toolmakers; we are *Homo faber*. Although other animals demonstrate the ability to use elements of their environments as simple tools, such as otters using stones to open the shells of mollusks, humankind is marked by its whole-scale commitment to develop and use new tools. It is hard for human beings to even image a world without the use of tools of some kind. Further, it is doubtful that as a species *Homo sapiens* would survive without tools, without technology.

Biotechnology is a set of technologies specifically aimed at manipulating living things, including human beings themselves, arguably for the common good. Some of the most amazing tools of the past fifty years have appeared in the sphere of medical biotechnology—

antibiotics, psychopharmaceuticals and other drugs, and recombinant technologies that produce hormones or clotting factors (e.g., insulin, erythropoeitin, or Factor VIII for hemophiliacs), gene manipulation that produces disease or pest resistance in crops, and engineered drugs to treat cancer (e.g., imatinib mesylate, or Gleevec, for chronic myelogenous leukemia, or rituximab for the treatment of lymphoma and an increasing number of autoimmune diseases). For the purposes of our discussion, biotechnology also includes tools that directly interact with the systems of the body for the purposes of diagnosis, health restoration, and disability amelioration. This category of biotechnologies includes organ and blood cell transplantation, pacemakers, new forms of orthopedic appliances, genetic testing and the first forays into gene therapy, neural implants to treat Parkinson's disease or depression, and a whole host of tools and treatments now common in medical practice.

Many of these advances have been welcomed enthusiastically, whereas others have been greeted with skepticism or open hostility (e.g., genetically modified food). Increasingly, discussions about the means of biotechnology have joined controversies about its ends. Witness the heated debate about embryonic stem cell research, or about human cloning with a goal of developing new treatments for serious diseases like Parkinson's disease, juvenile diabetes, and a host of others.

Increasingly, biotechnologies are being created and used not for therapeutic ends but for the purpose of "enhancing" mental or athletic function or altering physical appearance.[1] Breast implants and other cosmetic surgery, anabolic steroids and erythropoeitin, botulinum toxin (Botox), methylphenydate (Ritalin), and SSRI (selective serotonin reuptake inhibitors) antidepressants (e.g., Paxil, Prozac, and Celexa) are being used not for the purpose of healing or restoring but in the hope of making us "better than well."[2] In 2004, just under $12.5 billion was spent in the United States on cosmetic procedures (surgery and nonsurgical such as Botox injections), a value far greater than ten times that committed to research for a cure or more effective treatments for malaria, still one of the world's major killers.[3] New agents that are being developed to treat memory impairment in dementing diseases such as Alzheimer's disease are expected to have a huge market among those who simply want to improve their memory for socially competitive reasons.[4]

Revisiting Eugenics

One of the most remarkable biotechnology projects of the past fifteen years has been the discovery of the human genetic code, in the Human Genome Project (HGP). Yet much has been written about the potential sources of harm and benefit in this new knowledge about the code of life. Will we have to again endure the abuses of the past performed in the name of some eugenic ideal? We are already engaging in both positive and negative eugenics through the use of prenatal and preimplantation genetic diagnosis. Some are more concerned that the eugenic programs we face now and in the future will not so much be state sanctioned (although there are governments around the world explicitly promoting eugenic programs[5]) but rather will be coerced socially by a "eugenics of the marketplace."[6] Several authors have quite explicitly stated not only that eugenics in some form is inevitable but also that genetically reengineering the human species is desirable.[7] Thus one set of questions before us as the community of humankind, and as individuals, is: Should we deliberately reengineer human beings by genetic manipulation? Should we use our genetic knowledge to determine what kind of people will be allowed to exist?

Bypassing Genetics and Reproduction

Other technologies are being rapidly developed, however, that will enable human beings to reengineer themselves without the need to involve genetic and reproductive mechanisms. Rather, the existing person will soon be offered an array of means to remake himself or herself via tissue reconstruction or prosthetic enhancement. These additional means of human reengineering are cybernetics and nanotechnology. Because many readers may be less familiar with these fields than with the potential for genetic manipulation, we will present a brief overview of them and their potential for producing tools for reengineering humankind.[8]

Cybernetics

Cybernetics in its purest definition is the science of "control and communication in the animal and the machine" and was devised as a field by Norbert Weiner in the 1940s. The word is derived from the Greek for steersman, *kybernetes.* In *The Human Use of Human Beings,* Weiner wrote that "society can only be understood through a study of the messages and the communication facilities which belong to it; and that in the future development of these messages and communication facilities, messages between man and machines, between machines and man, and between machine and machine, are destined to play an ever-increasing part."[9]

In his *Introduction to Cybernetics,* W. Ross Ashby noted that this theory of machines focuses not on what a thing is but on what it does: "Cybernetics deals with all forms of behavior insofar as they are regular, or determinate, or reproducible. The materiality is irrelevant."[10] Recognizing that there are significant similarities in biological and mechanical systems, subsequent researchers have pursued the ideal of merging biological and mechanical/electrical systems into what Manfred Clynes and Nathan S. Kline termed *cyborgs* or *cybernetic organisms.*[11] In this sense, cybernetics has taken on the meaning of adding prostheses to the human, or animal, body to replace lost function or to augment biological activity.

Humans have long used tools to augment various functions, and for centuries they have also intimately attached some of these tools to their bodies. Filled or artificial teeth, glasses and contact lenses, hearing aids, pacemakers, and/or artificial limbs are all examples of this phenomenon. Recently, significant advances in the fields of neuroscience and computer technology have made possible the direct interface of animal or human nervous systems with electromechanical devices. A few examples in this evolving field are the creation of neural-silicon junctions involving transistors and neurons to prepare neuronal circuits, the remote control of mechanical manipulator arms by implants inserted into the motor cortex of owl monkeys, and remote controlling rats to move over a directed path via implanted electrodes and a computer moderated joystick.[12]

Investigators at the Max Planck Institute for Biological Cybernetics in Tübingen, Germany, have successfully grown connections between the neurons of several species of animals and transistors, allowing two-way communication through the silicon-neuronal junction. Researchers at Infineon Technologies in Munich, working in collaboration with the Max Planck group, announced the development of the "neurochip." This device has greater than 16,000 sensors per square millimeter and is able to record at least 2,000 readings per second (or in aggregate, 32 million information values per second). Neurons are kept alive in a special nutrient fluid that coats the chip's surface. The architecture of the chip ensures that each neuron in the matrix covers at least one sensor. Thus without the need to invade the structure of the neurons themselves, the chip can maintain prolonged undisturbed interactions between neurons and measure and process the flow of information through the neuronal network.[13]

Significant strides have also been made in understanding and manipulating the sense of vision. In 1999, Garrett B. Stanley and colleagues at the University of California, Berkeley, were able to measure the neuronal activity of the optical pathway of a cat via 177 electrical probes, process the information, and recreate rough images of what the cat's eyes were viewing at the moment. Though the images were of fairly poor resolution (similar to the degree of resolution of computed tomography [CT] scans in the early 1980s), it is only a matter of improved signal processing to be able to produce more exact images.[14] The implication of this line of research is that if we can decode the visual images from their neural representations, we will in time be equally capable of directly transmitting "visual" images into a recipient's brain, bypassing the use of, or need for, light-collecting visual organs.

Although this sort of technology could become the ultimate tool for virtual reality, the most important potential is for the restoration of sight to the blind.[15] Researchers at the Dobelle Institute, with its Artificial Vision for the Blind program (formerly headquartered in Zurich, with a laboratory in New York City; since 2001, headquartered in Lisbon) have implanted electrodes connected to a digital camera/computer complex into the visual cortex of blind patients, restoring some degree of sight. In one case, the patient was even able to drive around the parking lot of the hospital with some degree of restored vision.[16]

Investigators at Emory University in Atlanta have helped two patients with locked-in syndrome, a state in which the brain is conscious but cannot produce any movement of the patient's voluntary, skeletal muscles. The unfortunate patient is often thought to be in a persistent vegetative state. The two received brain implants, into which their neurons grew, establishing a link with a computer. This enabled the patients to use their minds to control a cursor on a computer screen and thus communicate with others.[17]

In June 2000, Chicago's Optobionics Corporation implanted the first artificial retinas made from silicon chips in the eyes of three blind patients suffering from retinitis pigmentosa. Each implant was 2 millimeters in diameter and 1/1,000 of an inch thick and contained approximately 3,500 microphotodiodes that converted light energy into electrical impulses to stimulate the remaining functional nerve cells of the retina.[18]

A team at the University of Southern California in Los Angeles is planning to replace part of the hippocampus of rats with a special silicon chip. The hippocampus is a crucial part of the brain where the laying down and retrieval of memories is coordinated. The implant would in effect be a memory chip.[19] Though designed as a treatment to replace damaged tissue in poststroke patients or patients with Alzheimer's disease, it is not difficult to imagine such a device (which at the time of this writing is estimated to be at least ten years off) being used for memory augmentation for the normal as well.[20] Though this is exciting on one level, devices such as this illustrate the significant technical and safety issues involved in developing and employing neuroprostheses. Even if such a device could communicate with the rest of the brain in storing memories, there are still serious issues to be resolved. Not only does the hippocampus participate in memory, but it is also a significant component of the brain pathways that affect awareness, consciousness, and emotions. Thus, how such a device may affect other significant elements of brain function and personality is unknown.

Medical Nanotechnology

As the name implies, *nanotechnology* is engineering or manipulating matter, and life, at nanometer scale—that is, one billionth of a meter. Ten hydrogen atoms side by side span 1 nanometer. The deoxyribonucleic acid (DNA) molecule is 2.3 nanometers in diameter. If such feats were possible, then it is conceivable that the structures of our bodies and our current tools could be significantly altered. In recent years, many governments around the world, including the United States with its National Nanotechnology Initiative, and scores of academic centers and corporations have committed increasing support for developing nanotechnology programs.[21] For example, in 2003 the U.S. Congress passed by a wide margin the Nanotechnology Research and Development Act of 2003 (it was signed by the president on December 3), authorizing that $2.36 billion be spent over three years on nanotechnology research.[22]

The idea behind nanotechnology originated with the physicist and Nobel laureate Richard Feynman (1918–88) in a speech given at an annual meeting of the American Physical Society at the California Institute of Technology in Pasadena on December 29, 1959. But the applications remained to be pursued into the 1980s. Feynman described the development of tools for molecular engineering—that is, building materials molecule by molecule. His startling claim was that this sort of task would not require a new understanding of physics but rather was completely compatible with what scientists already understood about the nature of fundamental forces and matter. When scientific and technological communities began to pursue Feynman's vision, Eric Drexler's works—which both demonstrated the feasibility of such manipulation from an engineering perspective and provided a vision for the possible benefits of such technologies—sparked a flurry of activity that continues to expand almost exponentially.[23]

The list of potential uses of nanotechnology continues to grow. An early focus of research has been in the area of miniaturization of electronic components,[24] but nanoscale materials may also dramatically improve the durability of materials used in machinery and potentially create production methods that will be less polluting and more efficient. The military has a significant interest in nanotechnology and has created the Center for Soldier Nanotechnologies (CSN).[25] Among the initial aims of the CSN are creating stealth garments (and coatings) that are difficult to see or detect, are highly durable, and may provide increased protection from penetrating objects; significantly reducing the weight of materials the individual must carry onto the field of battle; and developing devices to rapidly and accurately detect the presence of biological or chemical weapon attacks. The CSN is interested in the use of such technology to help create the seamless interface of electronic devices with the human nervous system, engineering the cyborg soldier.

Medical uses of microscopic, subcellular machines potentially include:

- rational drug-design devices specifically targeting and destroying tumor cells[26] or infectious agents,[27]
- in vivo devices for at-the-site-of-need drug manufacture and release,
- drug-delivery systems,[28]
- tissue engineering or reengineering,
- early detection or monitoring devices,
- in vitro laboratory-on-a-chip diagnostic tools,[29]
- devices to clear existing atherosclerotic lesions in coronary or cerebral arteries, and
- biomimetic nanostructures to repair or replace DNA or other organelles, provide artificial replacements for red blood cells and platelets,[30] augment or repair interaction between neurons in the brain, improve biocompatibility and the interface between brain tissue and cybernetic devices, and develop more durable prosthetic devices or implants.[31]

Such tools have also been envisioned to provide new means of cosmetic enhancement, such as new forms of weight control, changing hair or skin color, removing unwanted hair, or producing new hair simulations.[32]

The development of nanotechnology has taken two major pathways. The first is the so-called top-down approach, which attempts to directly manipulate matter atom-by-atom and molecule-by-molecule. This approach is heavily dependent on high-technology machinery such as atomic force microscopy. Other top-down approaches pursue a more traditional chemistry-based approach, for example, the production of carbon nanowires or carbon spheres, such as buckminsterfullerene or "bucky balls." Carbon nanowires have proved to be excellent conductors and are being used in a variety of ways to develop new forms of transistors and microscopic circuitry, leading to significant gains in the miniaturization of electronic devices. Bucky balls are being evaluated as the base for a variety of drug-delivery systems.

The second approach, the bottom-up approach, recognizes that the building blocks of life, all the enzymes and other components of each living cell, are already acting as little machines operating at the nanoscale. This approach then tries to use biological materials in new and different ways.[33] For example, a group of researchers led by Carlo Montemagno of Notre Dame University recognized that one of the critical enzymes in each cell in our body, a protein that is involved in the storage of energy in a molecule called ATP, actually rotates around a central axis like a motor during its function.[34] This same enzyme can work in reverse, taking ATP, releasing its stored energy, and producing rotary motion. The researchers developed a process to attach a small metal propeller to the central axis of this enzyme, called ATP-ase. They then exposed the new "molecular motors" to an energy source in the form of a solution of ATP molecules, just like what they might encounter in a living cell. Some of the motors spun the propellers around their axes. Though only a minority of the engineered molecules were able to perform this function (sometimes the propeller would fall off, etc.), some of the motors worked for several hours. Montemagno's team has now devised a way to couple these motors to photosynthetic molecules so that sunlight is converted into a steady source of ATP to fuel the molecular motors. His group is also working on ways of synthetically producing moving devices that propel themselves like amoebas.

Dan Shu and Peixuan Guo have noted that certain forms of viral ribonucleic acid (RNA) also bind ATP, releasing its energy and powering a DNA-packaging motor that enables the virus to be created with all the genetic material packed into the small capsule of each individual viron.[35] This system indicates the significant amount of mechanical force that can be generated and applied at the molecular level.

Another bottom-up technique has been to use DNA, RNA, and other biomolecules, the fundamental molecules in our genes and gene expression, as tools for computation or as structural components.[36] In 1994, Leonard Adleman at the University of California, Los Angeles, used DNA molecules to solve a complex mathematics problem known as the Hamiltonian Path Problem. Laura Landweber of Princeton University reported in 2000 that her team had made an RNA computer to solve a chess problem.[37] Others have used enzymes, normally involved in the production, encoding, and decoding of genetic material, to create analogs of digital logic devices, such as AND gates, OR gates, and NOR gates, similar to those that are used in the chips of calculators and computers.

Ehud Shapiro and colleagues at the Weizmann Institute of Science, Rehovot, Israel, have demonstrated that the DNA molecule also intrinsically stores sufficient energy to power calculations using these molecules.[38] Thus DNA provides not only the information but also the power. Shapiro believes that eventually (sometime in the next fifty years) biological devices will replace inorganic, silicon-based electronics. In another fascinating report, Nicholas Mano, Fei Mao, and Adam Heller have described the development of a biofuel cell that operates at body temperature and works off of our body's normal physiologic fluids. This could provide an intrinsic energy source for cybernetic devices.[39] Heller, of the University of Texas, Austin, invented this implantable biofuel cell.

Our cells also regularly produce structures that integrate inorganic metals into proteins. It has been shown to be quite feasible that DNA and cellular-controlled processes can be designed to construct molecules that will facilitate the biological–inorganic interface.[40] This would allow, for example, more seamless integration of our nervous system with electronic devices, furthering cybernetic developments. A recent report illustrates other uses for organic–inorganic melding. Gold nanoparticles were attached to DNA molecules, serving as antennas. Radio signal controls were then used to cause the DNA to unwind and start producing the protein encoded by the segment of DNA, producing in essence a radio-controlled switch for turning genes on and off. Further, by manipulating or reprogramming cells at the molecular and genetic levels, cells can also be turned into computers and factories, producing unique products or in vivo diagnostic and therapeutic devices that react dynamically in real time to the surrounding environment.

"Enhancement" and Technological "Immortality"

If many of the potential therapeutic uses listed above become reality, allowing more effective treatment of life's greatest killers, such as cancer, infectious disease, and vascular disease, it should be clear that longevity could be greatly increased. To the degree that these tools may be used to heal the diseased and disabled, we can rejoice. But there are those who predict that nanotechnology and cybernetics together will enable humankind to pursue a form of technological "immortality." Others, though looking forward to an increased life span, are more interested in "enhanced" function now.[41] Some individuals, calling themselves "transhumanists," explicitly promote the reengineering of humankind into some form or forms of "posthuman" beings.[42] Even the U.S. government has invested in a controversial project to reengineer human beings.[43] Yet even if not adopting such as extreme view or goal, it would seem that large numbers of individuals in the United States, and around the world, are enticed by all the potential technologies of "enhancement." The desires for modification may be rooted in wishes for fashioning oneself into a more socially acceptable image, attempting to improve self-esteem through reengineering, or making oneself more competitive in business, the professions, academia, or athletics. Unfortunately, the motivations behind these desires are usually socially driven fears, experiences of rejection or failure, or just plain greed, and they may reflect a social rather than biological pathology.

Evaluating Consequences

Technologies do not exist in social vacuums. There are strong social forces that motivate the development and use of each technology. Use of the technologies of "enhancement," or reengineering, would at first consideration seem to be something entirely individual. It might be said that individuals should have the freedom to do with themselves as they please. But each choice, each modification, may carry with it a significant social impact and create forces that take away the freedom of others.

Take, for example, the field of medicine somewhere between twenty years and forty years from now. Brain chips or wearable computers now make it possible for a physician to have immediate, twenty-four-hour/seven-day access to databases that contain all standard medical knowledge plus the latest medical information and practice guidelines. All ongoing clinical trials are immediately identifiable. This information might be obtainable via an external device, but this would require porting and might not be with the physician in the event of an emergency. Thus internal devices would be more efficient. Internal devices may also possess a variety of other desirable capabilities that wearables do not. For instance, the device may rehearse procedures in the visual and motor cortex, allowing physicians to maintain a high state of proficiency in a vast array of surgical procedures, even if not performed frequently. Indwelling nanodevices in the implant could provide pharmacological or direct neural stimulation to maintain alertness and calm. With the device, the cyborg physician might, by simply thinking it, summon colleagues for a consultation or call emergency personnel. If, as a patent, you were given a choice, would you want a physician with capabilities like this, or the standard nonaugmented model, dependent on a limited memory and personal experience and also potentially sleep deprived and grumpy? It is not difficult to imagine that there would be strong professional and social forces—not to mention huge differentials in malpractice premiums—compelling physicians to undergo cybernetic and nanotechnological modification. (If cyborg docs could be replaced by fully robotic nonhuman physicians, or *robo-docs*, then so much the better.) This could be true for many other professions as well, including law, engineering, finance, and the military.[44] Individuals may indeed dissent and decline technological augmentation, but such dissenters will find job options increasingly scarce. Moreover, the fundamentally human relationship between a physician and her patient will be altered radically by technology.

Because business networks, and essentially all systems of cyborgs, will require increasing levels of cooperation and harmonious coordination to improve the efficiency of the system and thus ensure the continued use of the prostheses, the devices themselves will be engineered or modified to introduce means of controlling or modulating emotion to promote these values. Meanwhile, also in the name of efficiency and societal harmony, the network will be increasingly controlled by central planning structures to facilitate these undeniably legitimate goals. Though everyone still considers himself or herself fully autonomous, in reality behavior increasingly will be controlled and modulated. Is this an unlikely science fiction scenario? Not really. Each of these technologies is being developed at the writing of this book. Our competitive, narcissistic society is already driving individuals to pursue any and all means to "get ahead."

The technologies of reengineering, and the question of how biotechnologies are developed and used, raise many concerns—concerns that require a clear understanding of the nature of human beings, our purposes for existence, and our personal, social, and environmental responsibilities. From these fundamental anthropological questions, we then must ask, what is the nature of technology? Are there appropriate goals and limits to the development and use of technology? What are the legitimate goals of medicine and medical technology? What should we expect from one another? Should we pursue the quest of "perfecting" ourselves when we have no real standard of what perfection would look like? Or should we simply allow market forces to regulate the ethics? With the costs of medical interventions for the purposes of healing already staggering, is it appropriate to develop and use devices that are solely for reengineering ourselves? Can we develop tools that have potential for healing but at the cost of destroying or demeaning other human beings in the process?

In the chapters that follow, our goal is to address these questions. We do so by following what we believe is a logical sequence of key questions needed to develop a framework for the evaluation of biotechnology, because each subsequent question depends on the answers to the preceding ones:

1. What is the nature of the human being?
2. What is the social nature and communal responsibility of human beings?
3. What is the nature of technology, including biotechnology?
4. What is the nature of medicine as a specific technology?
5. How then should we evaluate technologies, particularly biotechnologies?

The answers to these questions are necessarily based on a worldview, an understanding of the nature of reality, the context in which all these issues find their existence. It is only in carefully evaluating the many competing worldviews and concluding which is most consistent with the evidence that can we answer the major questions in such a way as to promote human flourishing. We therefore focus the next chapter on the critical foundation of worldview.

CHAPTER TWO

Humanity and the Technological Narrative

> The very identity of the human person and the very substance of reality are presumably called into question by developments in artificial intelligence, in genetics, and in virtual reality. Reactions to these prospects are as divided as they are to carnival rides—they produce exhilaration in some people and vertigo in others.
>
> —Albert Borgmann, "On the Blessings of Calamity and the Burdens of Good Fortune"

TECHNOLOGIES are *teleological.* That is to say, they have certain goals or purposes. *Teleologies* are value laden. Good ends are sometimes pursued, bad ends are sometimes pursued, and there is always the possibility that a technological aim is indifferent. Clearly, then, technology is not an unqualified good. This may come as no small surprise to our technologically saturated society. Many Westerners—and most North Americans—are not only technological optimists but also technological utopians. If a technology can help us perform a task faster, easier, and more powerfully, then most people believe it is necessarily a good thing. Yet because technologies, including biotechnologies, are value laden, they may be morally good, bad, or indifferent. So the values that shape, inform, and provide the impetus for technology must be examined.

Furthermore, the choice to develop certain technologies reveals a great deal about one's understandings of the purpose of a given technology. Technologies may also carry hidden costs, or as Tenner puts it so vividly, they sometimes "bite back."[1]

For instance, in the 1960s the snowmobile revolution took the Arctic by storm. The Skolt Lapps of northeastern Finland had made their living for centuries by herding reindeer. After World War II, the Skolts were given a choice either to return to rule by the Soviet Union or to come under the Finnish government. They chose the latter. By the late 1950s, about fifty Skolt households survived, consisting largely of nuclear families still loyal to the Russian Orthodox Church. In the prewar days, the Skolts relied mainly on fishing for their food supply. Reindeer were a secondary food supply. Not only so, however, the reindeer were also the principal means of income. Reindeer meat, hides, and other parts were traded for cash to buy the modern necessities such as flour, sugar, coffee, tea, and nonfood commodities such as medicine. The ethnographer Pertti J. Pelto observes that "as long as a man had reindeer herds he could always get food for his family, and he could also sell animals to get needed cash."[2]

Before the introduction of the snowmobile, reindeer had been gathered in domesticated herds grazing pastorally in the Arctic tundra (if "pastorally" and "tundra" are not an oxymoron). In 1960, there were forty-one active herders. Throughout the next decade, the technologization of reindeer herding radically altered Skolt society. As the snowmobile became the preferred method of herding, the reindeer were de-domesticated. Because the snowmobile provided a fast and efficient means for gathering the animals, they could be allowed to run wild until it became necessary to round them up. De-domestication came with certain costs to the animals, however. Because herding changed from a peaceful process to a chaotic rampage, the reindeer experienced an increased incidence of lung damage as they bolted from the growling snowmobiles. Also, the shift from pastoral herding to mechanical round-ups lowered the weight of the reindeer population; this, in turn, resulted in low-birthweight calves and a less healthy herd.

Moreover, the snowmobile revolution resulted in economic depression and socioeconomic stratification. As the costs of herding escalated with the costs of the technology and the petroleum products required to operate the technology, fewer Skolts could afford to herd reindeer. The cost to produce one animal rose from $1.30 in 1963 to $3.30 in 1969. Only one-third of herders in 1960 were still herding in 1971.

Pelto opines that "the advent of mechanized herding had created a situation of 'technological unemployment' for the men who were forced out of full-time herding activities."[3]

Atlhough this story reveals the downside of technology, hundreds of thousands of stories could be told of how technology has revolutionized society for good. From simple tools to space exploration, some technologies have improved human, animal, and other creatures' quality of life.

In this chapter, we explore the development of technologies as a uniquely human activity rooted in our own understanding of our role in the world. The way humans understand their role(s) in the world underwrites the impetus for technological expansion—it reveals their understanding of the teleology of technology. Especially in the American context, a Judeo-Christian worldview has provided the construct for the development and use of technology. However, even among Jewish and Christian theists, there is more than one way of understanding the role of human technological expansion. Here we elucidate several possible ways of understanding our role and argue for a model of responsible technological stewardship.

The Role of Technology's Foundation Stories

In *America as Second Creation: Technologies and Narratives of New Beginnings*, David E. Nye, professor of American Studies at Odense University in Denmark, explores the role of technology through the lens of what he calls *foundation stories.* These narratives provide a useful and illuminating means of analyzing the teleology of technology. "Americans," says Nye, "constructed technological foundation stories primarily to explain their place in the New World."[4] He argues that Americans saw this new world as an opportunity for a second creation. In fact, they saw *re*-creation as a moral imperative and as necessary to their existence.

The Second-Creation Narrative

Though invoked at different times and on varied occasions, second-creation technological foundation stories participate in a similar structure, as follows:

- A group (or an individual) enters an undeveloped region.
- They have one or more new technologies.
- Using the new technologies, they transform a part of the region.
- The new settlement prospers, and more settlers arrive.
- Land values increase, and some settlers become wealthy.
- The original landscape disappears and is replaced by a second creation largely shaped by the new technology.

The process begins again as some members of the community depart for another undeveloped region.[5]

This is largely the outline of the story of the American technological founding (and, we might add, the story of contemporary biotechnology). The story has, according to Nye, four acts. The first act is the narrative of the axe. James Fenimore Cooper famously exclaimed:

> The American axe! It has made more real and lasting conquests than the sword of any warlike people that ever lived, but they have been conquests that have left civilization in their train instead of havoc and desolation.[6]

Or as Timothy Walker put it in the *North American Review* in 1831:

> Where she [nature] denied us rivers, Mechanism has supplied them. Where she has left our planet uncomfortably rough, Mechanism has applied the roller. Where the mountains have been found in the way, Mechanism has boldly leveled or cut through them.[7]

The axe permitted mastery of the unimproved natural resources. With each fateful blow, the untamed wilderness was brought into subjection, rendering it domesticated and inhabitable, not to mention marketable. The prevailing notion of property acquisition at our nation's founding was John Locke's theory of the labor acquisition of property:

> Before the Appropriation of Land, he who gathered as much of the wild Fruit, killed, caught, or tamed, as many of the Beasts as he could; he that so employed his Pains about any of the spontaneous Products of Nature, as any way to alter them, from the state which Nature put them in, by placing any of his Labour on them, did thereby *acquire a Property in them.*[8]

For Locke, because the body is one's property and because the employment of labor is an extension of one's body, the commons become personal property when one's labor is used to tame nature.[9] And our forebears did tame nature.

According to Nye, the second act of America's foundation story is the era of the water-driven mill. The mill was another method of harnessing the power of nature to establish the second creation. And here the efficiency motif is clearly in evidence. For instance, as the award-winning historian of technology Eugene Ferguson notes:

> To grind a bushel of wheat to flour requires about two horsepower-hours of work, the equivalent of two man-days of strenuous effort. In a horse-drawn mill, the process would require a third of a day, while a water mill of moderate size might produce fifty bushels of flour a day.[10]

The third act in the second-creation narrative, according to Nye, was the era of the canal and railroad. Mechanized transportation conquered the problem of space. Technological expansion was not without its critics. As early as 1839, Ralph Waldo Emerson wrote in his journal: "This invasion of Nature by Trade with its Money, its Credit, its Steam, its Railroad, threatens to upset the balance of man, & establish a new Universal Monarchy more tyrannical than Babylon or Rome."[11] Nevertheless, the technocrats won the day with then–House of Representatives member John C. Calhoun's cry:

> Let us . . . bind the Republic together with a perfect system of roads and canals. Let us conquer space.[12]

The final act in America's foundation story was irrigation, according to Nye. Irrigation made it possible to declare independence from the fickle clouds by bringing the rivers into servitude to fuel the factories in the field. The U.S. Reclamation Bureau built dams, canals, and irrigation systems with a view toward turning arid lands into lush Edens. Paradise would be regained.

Nye does a masterful job showing how second-creation stories pervaded the American sense of the past by 1900. Obviously, competing stories were told, but they by no means captured the American imagination the way the second-creation story did.

Another social historian and philosopher of technology provides a similar account of the evolution of technology. In his monumental study *Technics and Civilization*, Lewis Mumford chronicles four stages of human technological invention.[13] Each stage represents phases in the effort to exploit raw materials and natural resources in the development of the so-called machine age. The first stage was what Mumford called the age of *eotechnics*—"the dawn of the age of modern technics."[14] Eotechnics embodied the era of water and wood, when human production consisted in the manipulation of the forces of nature. Windmills and waterwheels harnessed the energy that had been the object of ancient veneration—wind and water. Wood was used for building the eotechnic era but, more important, during this period all three (wind, water, and wood) were combined to manufacture sailing vessels.

The next phase, the *paleotechnic* phase, resulted in the Industrial Revolution. Coal and iron were employed to produce the energy requisite for mass production. Mumford maintained that the paleotechnic age reached it apogee in "the great industrial exhibition in the new Crystal Palace at Hyde Park in 1851: the first World Exposition, an apparent victory for free trade, free enterprise, free invention, and free access to the world's markets by the country that boasted already that it was the workshop of the world."[15]

The third stage was the *neotechnic* era. During this phase in the evolution of technology, electricity became a new form of energy. At the same time, the neotechnic age saw the invention of a host of new synthetic compounds, including celluloid, Bakelite, and synthetic resins. Furthermore, the high level of conductivity required by the use of electricity led to the exploitation of copper and aluminum. The neotechnic age also celebrated the invention of the internal combustion engine, a revolutionary innovation indeed.

Finally, and most important for our purposes, Mumford pointed to what he called the "*biotechnic* period, already visible on the horizon" in 1934, when he wrote *Technics and Civilization*. This stage would be characterized, he prophesied, by a "completer integration of the machine with human needs and desires."[16]

Mumford further argued that the machine "devaluated" rarity (because machines could produce a million copies of the master model), age (because machines placed emphasis on adaptation and innovation), and archaic taste (because machines established new standards of value). The upshot of the evolution of the machine, for Mumford, meant that "we cannot intelligently accept the practical benefits of the machine without accepting its moral imperatives and aesthetic forms."[17]

Mumford may well have been playing the role of cultural prophet when he said presciently that "technics, instead of benefiting by its abstraction from life, will benefit even more greatly by its integration with it."[18] The emergence of biotechnology is a case in point. Animated by both the machine metaphor and the second-creation narrative, biotechnology now seeks mastery over our own biology—bringing it into submission, exercising labor-added property rights, and the like—in the effort to remake the natural order and subdue it for our own ends.

The Recovery Narrative

The second-creation story is not the only way to construe our technological relationship to the natural world. The recovery narrative offers a competing vision. Rather than reject the technological creation story, the recovery narrative emphasizes the spoiling of the landscape and the corruption of natural resources through arrogant exploitation. According to this account, notes Nye, selfish individualists have imposed free-market values on creation for short-term gain.[19] Pollution, erosion, degradation, and misuse constitute the storyline of this narrative. Consequently, the recovery story emphasizes initiatives to repair, preserve, and repristinate nature. Says Nye: "The process began with the preservation of Yosemite in 1864 and Yellowstone in 1871 and continued with the establishment of state and national forests."[20] The decommercialization of Niagara Falls and the turning of its shores into national parks exemplify the direction of the recovery narrative during the twentieth century. This is a narrative characterized by federal laws, environmental regulations, and managed preservation.

The Wilderness Tale

A final narrative describing the American technological experience is the wilderness tale, argues Nye. "If the conservation movement rewrote second creation into a recovery narrative, the wilderness movement wanted to keep some portion of nature pure and prevent it from becoming a part of *any* human story" (emphasis added).[21] In this narrative, humans are never the hosts, they are always the guests of nature; interlopers, if you will. Wilderness is always the foil to civilization and exploitation. In this narrative, pristine wilderness is viewed as more perfect than any human-made so-called improvement. Rather than imposing human ideals on the back of nature, "the wilderness story [is] about ecologists who [try] to think like a mountain."[22] Nye gives an example of that form of storytelling when he appeals to authors such as Jim O'Brien:

> Imagine the settlement of the United States from the viewpoint of animals. Jim O'Brien's often-reprinted essay "A Beaver's Perspective on North American History" begins before any humans had arrived. Beavers dammed streams, created small canals and ponds and reshaped the landscape in ways that affected the larger biotic community. The beaver population, which may have reached 60 million at its height, went into decline with the arrival of the European fur trade. O'Brien's narrative radically reconsiders American history as a catastrophe.[23]

For many who read the story of universe through the lens of the wilderness narrative, human beings are either mere appendages to the created order or, worse, enemies. We are either guests or invaders, not hosts and caregivers. The order itself is exalted as perfect, a living organism, which if only left alone would continue its own glorious existence. Though this narrative may offer attractive features, it seems to ignore or at least underestimate the reality of a fallen order where the world is not self-preserving and self-improving.

Shortcomings of the Three Foundation Narratives

Each of these narratives seems problematic to us. For example, the second-creation narrative sees the role of technology as characterized by mastery through the application of force for the purpose of human aggrandizement. The recovery narrative, though laudable in some respects, risks offering too little too late. Some natural resources, once squandered, may not be recoverable. Moreover, the arrogant exploitation of some resources (e.g., the atmosphere) might be lethal. The wilderness narrative seems to lead to the conclusion that nature, if left unspoiled, is already perfect. This means ignoring natural disasters like tornados, volcanoes, and droughts. Pristine wilderness is hardly bloodless and genteel. The law of tooth and talon reigns in places where the beauty of the wild masks its ferocity. Surely, even the most idyllic wilderness is not the best of all possible worlds.

Responsible Technological Stewardship

Over against these three narratives, we pose yet a fourth narrative: *responsible technological stewardship*, grounded in a biblical-theological understanding of human relationship with God, with the creation, and with one's neighbors.

What is technology? The economist John Kenneth Galbraith has defined technology as "the application of organized knowledge."[24] Though that expression may define technology in its very broadest terms, as a definition it seems less than illuminating. The Calvin College political science professor Steven V. Monsma offers a very helpful definition when he and his colleagues say, "We can define technology as a distinct human cultural activity in which humans exercise freedom and responsibility in response to God by forming and transforming the natural creation, with the aid of tools and procedures, for practical ends and purposes."[25]

Five elements inform Monsma's definition. Technology is, first, an activity. It is the activity of *Homo faber* (humans, the makers or fabricators). Moreover, it is an activity of *Homo faber* in community with others. This emphasizes among other things that individuals are in relation to other individuals. Technology is not autonomous, and neither are technologists.[26] The teleology of technological development must take neighbors into account.

Second, in Monsma's definition, technology involves persons exercising freedom and responsibility in response to God. This notion is set against technological determinism. We are not *governed* by technologies; we make *choices* about technology. To quote an earlier work by Nye:

> Machines are not like meteors that come unbidden from outside and have an "impact." Rather, human beings make choices when inventing, marketing, and using a new device.[27]

These choices are informed by our values, our notions of right and wrong, our understandings of justice, and our social preferences.

Third, according to Monsma, technology is a transformational project. We apply our choices to the environment around us, both organic and inert, both living and nonliving. We form existing objects, and we transform or manipulate objects from one form to another. Technological activity is not meant to sustain status quo but rather to alter our environment in one way or another.

Tools and procedures are, fourth, the means by which we form and transform. In this definition, technology is not first an object but an activity. Granted, technological artifacts result from this activity, but technology is not a "thing"—it is a behavior, the application of which uses tools and processes.

Fifth and finally, in Monsma's definition, the tools and procedures are aimed at practical ends. This feature separates technology from art. Utility is important to technology. Making and doing are aimed at achieving some practical goal. Art serves different ends.

The most important aspect of Monsma's definition, vis-à-vis other definitions, is its theocentricity. Because technology is the exercise of human freedom and human responsibility in relation *to God*, its application requires a careful stewardship. Responsible stewardship is the teleology of technology. Stewardship implies accountability. Accountability implies that something is "given" for which one stands accountable as a steward or caregiver.

Technology and Biblical Anthropology

The notion of responsible technology is consistent with a biblical anthropology of technology. Let us trace the lineaments of such an anthropology. First, God made the world (Gen. 1:1). Ultimately, everything is preowned by the sovereign Creator. Because everything is created by God—from light to darkness to animals to humankind—everything belongs to Him. As the Psalmist says: "The earth is the Lord's and the fullness thereof, the world and those who dwell therein, for he has founded it upon the seas and established it upon the rivers" (Ps. 24:1, New Revised Standard Version translation). Furthermore, from the beginning, humankind was commanded to "be fruitful and multiply and fill the earth and subdue it and have dominion over the fish of the sea and over the birds of the heavens and over every thing that moves on the earth" (Gen. 1:28). Again, as the Psalmist exclaims, "You have given [humankind] dominion over the works of your hands; you have put all things under his feet" (Ps. 8:6).

We are told in Genesis something about the character of this dominion when humankind was given the command to tend the Garden: "The Lord God took the man and put him in the Garden of Eden to work it and keep it" (Gen. 2:15). *To work it and keep it.* The Hebrew word translated "work" here is easy enough to understand; it is *abad*. It is sometimes translated as "to till" (Gen. 3:23, 4:2, 12). "Keep it" is the word *shamar*, and it specifies the nature of Adam's labor. Interestingly, the word is used of the occupation of Abel (Gen. 4:9), who cared for the land and his flocks. The word is also used (in Gen. 28:15, 20) of protecting persons. When God appeared in Jacob's dream, He said he would "keep him" wherever he went. Finally, the word is used of the priests who "served" God, faithfully carrying out His instructions (Lev. 8:35; Num. 1:53, 18:5). Thus, stewardship over the creation militates against worshiping the creation, and accountability to the Creator serves to check against abusing the creation. Both responses to creation are to be avoided. Stewardship sees to it that they are.

Furthermore, though the exact boundaries are not drawn for us, we can infer from the Creation account that some knowledge is meant to be possessed by stewards and some is not:

> The Lord God took the man and put him in the Garden of Eden to work it and keep it. And the Lord God commanded the man, saying, "You may surely eat of every tree of the Garden, but of the tree of the knowledge of good and evil you shall not eat, for in the day that you eat of it you shall surely die." (Gen. 2:15–17)

This text also demonstrates that the application of technology was meant to be a feature of human stewardship. Working and tilling the Garden required tools. By Genesis 4:22, we know that tool making was a recognized vocation, for Zillah was "the forger of all instruments of bronze and iron." Not only were tools a feature of early human life, even in the Garden of Eden, but procedures of classification were also utilized. Adam named all the animals, a process that can be taken to be a form of taxonomic classification (Gen. 2:20).

Alas, the created order did not remain Edenic. Adam and Eve acted irresponsibly by disobeying God's command. Thus a divine curse on the entire order followed human sin. God declared to Adam: "Cursed is the ground because of you; in pain you shall eat of it all the days of your life; thorns and thistles it shall bring forth for you; and you shall eat the plants of the field" (Gen. 3:17b–18). In eschatological hope, the Apostle Paul adds: "For we know that the whole creation has been groaning together in the pains of childbirth until now" (Rom. 8:22). Human sin and its curse wreaked havoc in the universe. Creation is not pristine, therefore. Entropy prevails for now. Because of the fallenness of the world, we should not assume its indestructibility; and because of the eschatological hope, we should not pander to its fragility.

Boat building (Gen. 6:14 ff), weapon making (Gen. 10:9), and building construction and city planning (Gen. 10:20) were early technologies in the lives and work of the people of God. *Homo sapiens* (human knowers) were also meant to be *Homo faber* (human makers). As Petroski expresses it, "To engineer is human."[28] To create, however, is divine.

Where human craft flourishes, however, human pride is not far behind. In Genesis 11, we have the account of the technological imperative run amok. As the people settled new territories, their hubris grew. Finally they cried: "Come, let us build ourselves a city and tower with its top in the heavens and let us make a name for ourselves" (Gen. 11:4). So they used their human technological skills to build monuments to themselves. The theologian of technology Graham R. Houston says of the towers of Babel:

> Excavated inscriptions indicate that these towers were meant to serve as stairways to heaven. They had a purely religious significance and had no practical use apart from religious ritual. According to the biblical narrative, they were symbolic of the desire to usurp the authority of the landlord. They were declarations of independence from the true God, yet also expressions of underlying religious needs.[29]

The Risk of Irresponsible Innovation

Sadly, the story of human technological innovation is not always a story of liberation. As Monsma points out so clearly, "When human beings set themselves up as masters of their fate, they set themselves up not for an ascent to freedom, as they imagine, but for a descent into slavery."[30] Technology unfettered from moral responsibility becomes the worst sort of tyranny. Or as C. S. Lewis so wonderfully puts it in *The Abolition of Man*, "What we call man's power over nature turns out to be a power exercised by some men over other men with nature as its instrument."[31]

Not only so, but to be human is to be subject to error.[32] Airline crashes, automobile tire failures, and the kudzu vine are troublesome reminders of the frailty of *Homo faber* and the fact that omniscience is not one of the communicable attributes of God.[33] Humankind's biggest problem is not a lack of ingenuity but a lack of responsible stewardship that attends to our own fallibility and that respects the propensity of technological artifacts to misfire.

The Prophetic Witness of Responsible Biotechnology

"The cadence of our culture is set by the beat of the technological drum," as Monsma puts it.[34] There can be little doubt that contemporary culture pulsates to the rhythm of technology. Technological development does not proceed in a normative vacuum, however. Choices about technology are value laden. Responsible technology must be directed to take these normative values seriously and to anticipate where decisions about technological development might lead.

The Skolt Lapps are a case in point. A new technology—the snowmobile—was available. This technology promised greater efficiency; that was its teleology. Greater efficiency must (or so people thought) translate into greater productivity. Greater productivity must translate into greater income. Greater income must translate into a better quality of life. Alas, would that it were so. In fact, as we saw above, the introduction of the snowmobile had tragic effects on Lapp culture, leading effectively to its demise.[35]

The story of biotechnology will be even more dramatic, to be sure; for clearly some biotechnologies are qualitatively different from others. It is one thing to alter form and transform the natural world around us; as chapter 6 demonstrates ahead, it is quite another to seek to form and transform the biology of the transformer. That is to say, it is a gargantuan step to move from altering the natural world to altering humankind through biotechnologies. If the effect of technological change can eradicate entire cultures like the Skolts', what might be the results when it targets the human species?

The answer depends in part on the foundational narrative to which one appeals. If one adopts the second-creation narrative, then *remaking* humanity can be justified as a form of *co-creation*. Humankind is warranted in altering itself because creativity is a God-like activity. Ronald Cole-Turner seems to be an exemplar of this approach. In *The New Genesis: Theology and the Genetic Revolution* he says:

> Genetic engineering does not encroach upon the scope of divine activity. It expands the reach of God's action, placing a new mode of contact, through our technology, between the Creator and the creation. God now has more ways to create, to redeem, and to bring creation to fulfillment and harmony. Human beings who seek to serve God through genetic engineering are placing new instruments, namely, their technical skill, into the hands of the Creator.
>
> . . . Long before our arrival on earth, God the Creator was at work through microbiological processes. Through billions of years of creation, God has opened up the creation for a stunning diversity of species. Working through untold instances of mutation, recombination, and natural

> selection, God has drawn out the creation patiently, step by step. Only in the most recent moment of creation have we appeared, and already our technology is giving us the power to add to this great work of creation.[36]

Cole-Turner's narrative represents the triumphalism of a second-creation story. But even on this hypothesis, who is to say that the appearance of humans extends God's great work of creation? If the evolutionary watchmaker is blind, who is to say that he did not leave out a few parts here and there?[37] That is to say, one's assumptions (or conceptual narrative) make a difference in how one conceives of one's technological role in the world and, therefore, the teleology of technology.

Or take the transhumanists as an example of second-creation storytellers. Kevin Warwick, professor of cybernetics at the University of Reading in England, argues that he deserves the "distinction" of being the world's first "cyborg" because he wears implanted computer chips in his arm and wrist.[38] The next stage of human evolution, says Warwick, is the cybernetic age. As he told *Newsweek* magazine in January 2001:

> The potential for humans, if we stick to our present physical form, is pretty limited. . . . The opportunity for me to become a cyborg is extremely exciting. I can't wait to get on with it.[39]

Conclusion

When we enter into the story of responsible technological stewardship, we understand ourselves as caretakers who must give an account to the Owner for the ways we have preserved, and even improved, what we have been given. Technology and technological development must be respectfully harnessed and carefully directed to avoid harm, and they must be wisely shepherded to achieve appropriate goods. Biotechnology, as a species of this larger project of technological development, must advance, but not without careful planning. In fact, just as the Judeo-Christian worldview provided the basis for modern science, so it provides the impetus for responsible technology.[40] At the same time, the worldview we inhabit, along with its narrative, provides the modesty for caution and the mature knowledge that if we are not extraordinarily careful we might cause more problems than we fix with technology.

Furthermore, the responsible stewardship narrative will not admit technological determinism. We are not pawns in some vast technological game. We make choices about technology—choices for which we must give an account. We make choices about technology—choices for which our children and their children may suffer.

Having said that, we do believe that some technologies, once chosen, overcome inertia and create a certain momentum that may not be altered for generations. For instance, under President Dwight Eisenhower in the 1950s, Americans decided to invest in a system of interstate highways. Once that decision was made, it set wheels in motion (no pun intended) that continue to this day. Now, more than a half century later, the consequences of our decision includes both benefits and burdens. For instance, because the highways permit interstate trucking, most of us enjoy fresh produce year round. But it is also the case that nearly every American owns at least one car, which he or she drives daily. It is also true that we face unprecedented air pollution. It is also true that we cannot survive (at least currently) without massive amounts of petroleum.

Conversely, in the United Kingdom and in many parts of continental Europe, the decision was made to invest more in public transportation than in interstate highways. Consequently, fewer people own cars, public means of transportation are more ubiquitous, and less fossil fuel is required. Admittedly, trains also sometimes fail to arrive at their destinations on time, may be overcrowded, and, like automobiles, sometimes crash.

Our point is not to recommend one choice over the other but rather to point out that the choice to invest in one technology over another produced a momentum that future generations have either enjoyed or endured. Biotechnologies, especially species-altering technologies, are even more potent than transportation technology. And as Bill Joy, the founder and former chief scientist at Sun Microsystems (in Santa Clara in the Silicon Valley), and others worry, these technologies are now available to far more people and groups than some earlier forms of technology.[41] The choices we make about the use of technology to alter our own humanity are decisions for which we will one day give an account. We will be held responsible. Future generations will look back on our choices. Will they bless or curse us? Even more sobering, God will hold stewards accountable for their choices. Will God reward us for our careful stewardship or charge us with reckless dereliction of duty?

CHAPTER THREE

Biotechnology and Competing Worldviews

> Those who have not discovered that worldview is the most important thing about a man, as about the men composing a culture, should consider the train of circumstances which have with perfect logic proceeded from this. The denial of universals carries with it the denial of everything transcending experience. The denial of everything transcending experience means inevitably—though ways are found to hedge on this—the denial of truth.
>
> —Richard Weaver, *Ideas Have Consequences*

BIOTECHNOLOGY: The term conjures up visions of science fiction to some, and to others it suggests the ultimate postmodern hope for human beings to remake themselves according to their own design. Others are cautiously optimistic about the benefits of the emerging biotechnologies, seeing the potential for good but recognizing the prospects for creating a world in which we do not want to live. But still others, such as William Kristol and Eric Cohen, suggest that biotechnology is moving quickly ahead without much consideration for what type of society might result.[1] Biotechnology raises not only ethical questions but also broader and more profound philosophical questions—about the goals of medicine, about human nature, about the nature and purpose of technology, and ultimately about one's view of the world. Kristol and Cohen assert that "the debate over genetics" (and also, we would add, broadly over biotechnology) "is about whether or not man and women should remake, redesign, and prefabricate themselves and their offspring, and about whether these new genetic powers will lead, quickly or eventually, to the 'abolition of man' "—in C. S. Lewis's words.[2] Similarly, the biotechnology critic Jeremy Rifkin argues that worldview components such as cosmology and one's view of nature are critical to the newly emerging intersection of biotechnology and economics.[3] He suggests what we must ask ourselves:

> The role cosmology plays in rationalizing the new economic circumstances society finds itself in is critical. . . . Concepts of nature focus on the big questions: Where did we come from? Why are we here? Where are we headed? . . . The fact is, we human beings cannot live without some agreed-upon idea of what nature and life are all about.[4]

He further points out that today's biotechnological changes are "being accompanied by a revised cosmological narrative."[5]

The general public often supposes that the debate over biotechnology concerns whether or not specific technologies ought to be developed and made available, either through government or through the market. But often not noticed in these debates are the profound assumptions that are rarely exposed or questioned, about how people see the world and how their view of biotechnology fits, or does not fit, with their worldview. One glaring example of this comes from the unabashed biotechnology proponent Gregory Stock, director of the Program in Medicine, Technology, and Society at the University of California, Los Angeles, who laments that those who oppose or question aspects of biotechnology represent "people whose worldviews in this sensitive area hold *an inordinate sway* on their thinking" (emphasis added).[6] Of course, Stock does not realize that his own worldview holds great sway on his thinking (as it does for everyone involved in biotechnology); and the assumptions in his worldview are likely unstated and unexamined. And that is our purpose here: to analyze the competing worldviews, or "faiths" or belief systems, held both by those who advocate and by those who are critical of the emerging biotechnological revolution.

What are the different faiths or sets of assumptions vying for influence in biotechnology today?[7] Although it is impossible to account for all the variations in how people see the world, we can distinguish three major players in the biotechnology arena. At one end of the spectrum, most advocates of biotechnology are what are known as *philosophical naturalists*. Sometimes referred to as *scientific materialists*, adherents to this worldview are by far the most influential in determining the direction that biotechnology will take. Most philosophical naturalists hold that reality consists of the material world alone, that what is real is what is empirically verifiable.[8] Naturalists hold to a view of a human person known as *physicalism*, asserting that a person is no more than his or her biological parts. They have allies in what we will call the *pragmatic libertarians*. This group suggests that the only constraints biotechnology should be subject to are those of the market. They are libertarian in their political philosophy and pragmatic in weighing the costs and benefits of biotechnology. Their view of a good society is what drives their view of biotechnology.

At the other end of the spectrum, rejecting virtually all biotechnologies, are the radical environmentalists, a small minority who hold that the natural world is virtually sacred and should not be altered by human intervention. This view is commonly known as *biocentrism*, what we will refer to as *environmental biocentrism*, and is indebted to the wildness narrative discussed in the previous chapter. This group opposes the application of biotechnology to human beings, most genetic technologies, and genetically altered foods.

A third group, one that is enthusiastic about responsible biotechnology, holds to *Christian theism*. This group strongly supports technology that alleviates the effects of the entrance of sin into the world (e.g., disease and physical deterioration), but it has concerns about violations of human dignity when it comes to biotechnological enhancement or more radical notions of remaking human nature. This book is about articulating and applying this type of Christian worldview to biotechnology. Such a view of the world will have a good deal in common with other religious worldviews, such as Judaism and perhaps even Islam.

This chapter introduces the key elements of a Christian worldview and argues that it is the only worldview with the resources adequate for grounding human dignity and ensuring that biotechnology develops within proper parameters. We focus our scrutiny on the major player in biotechnology, the prevailing philosophical naturalism of the scientific/technological community, and hold up beside it a fuller alternative worldview grounded in Christian theism.

Defining a "Worldview"

Everyone has a worldview, whether one realizes it or not; or whether it is well considered or not. A worldview is a set of beliefs or assumptions through which one sees and evaluates the world. It is a set of fundamental commitments that a person has about the world and the way it works. Every worldview must answer a set of critical questions, including the following:[9]

1. What is the nature of reality, especially what is ultimate reality? (These are questions of metaphysics.)
2. How can we know the world? That is, what do we know and how do we know what we know? (These are questions of epistemology.)
3. What is human nature, or what is a human being? (These are questions of anthropology.)
4. What is right and wrong and on what basis do you make moral decisions? (These are questions of ethics and morality.)
5. What happens to a person at death?
6. Where is history going, or what is your view of history?

How one answers these questions determines one's worldview. Of course, not everyone has a well-thought-out view of the world, nor does everyone have a worldview that is entirely consistent. Most people have simply not thought that much about how they view the world. In our culture, it seems that thinking through one's worldview is not highly valued. But for the Christian, who is called to "love God with all your mind" (Matt. 22:37) and "to be ready to present a defense for the hope within you" (1 Pet. 3:15), thinking hard about your worldview is a critical component of your relationship with God and your effectiveness as an agent of cultural change.

The Worldview of Philosophical Naturalism

There is little doubt that in the academic world, and especially in the world of the sciences, the dominant worldview is philosophical naturalism, or scientific materialism. That is not to say that there are not theists among scientists. Clearly that is not true, because many scientists have a real and vibrant faith. But for many believers in the sciences, their faith is a private matter and has little impact on their work in science. For some, their faith has an impact on their work and how they view it. But the vast majority of the scientific community has embraced the worldview of naturalism, with all that it implies for their work.

What is meant by the term *philosophical naturalism*? What are the components of such a worldview? Essentially, it means that all reality is subsumed in the material world alone—that is, what can be appraised by one's senses and what can be empirically measured and verified. There is no ultimate reality beyond this material/physical world. In terms of metaphysics, there is nothing that is "meta" or beyond the physical world. To put it another way, the universe that exists in space and time, which consists of physical objects, physical properties, events, and process, is all there is, ever was, and ever will be. As a result, there is no room for immaterial human components that cannot ultimately be reduced to physical or material substances. For example, consciousness and self-awareness must be capable of being reduced to the neurological components in the brain or the behavior of the body. There is no room for spiritual and nonmaterial notions such as the soul and the image of God.

Naturalism also contains a story about how reality came to be. Naturalistic evolution is the "myth," or story, about the origin of reality and a statement that ultimate reality is reducible to the physical world. In this view, the universe is a closed system, and all events that happen in it are the result of past events and the laws of nature.

Epistemologically, all that can be known is what is scientifically verifiable, and such things are the only things that count as knowledge. This epistemology is known as *scientism.* What cannot be verified is counted as mere belief and is often in sharp contrast to knowledge. Scientism has established an epistemological hierarchy in which "scientific" knowledge clearly is either the only kind of knowledge we have (actually *epistemological reductionism*) or is vastly superior to non-"scientific" claims, which amount to private expressions of belief. Knowledge is publicly accessible, verifiable, and objective, and it can make a claim to be truth (contrary to the postmodern incursion into the hard sciences). According to the naturalist, belief is private and subjective, cannot be tested, is ultimately a matter of personal preference, and can make no valid claim to truth or authority over someone else.

Another critical component of naturalism is its view of the human being. According to the naturalist, human beings and all life on earth are the result of the blind, random forces of evolution. There is no design or intelligence behind the world and no purpose or reason behind it that would invest life with significance. Human beings are the end product of billions of years of evolutionary change, and according to many adherents of biotechnology, human beings will likely be surpassed by a biotechnological creation in the future. As a result of the evolutionary story of origins, there is no transcendent source of life's meaning or of human dignity. Both of these are human creations, according to the materialist worldview, which today sometimes takes a postmodern shape, suggesting that notions of human dignity are fluid and changing as they reflect culture and power relationships. For the strict naturalist, no basis exists for human dignity or life's significance because all life is the result of random forces. The naturalist tends to view human beings in a deterministic way, rendering free will an illusion. The view of ethics and morality is very similar in its assumptions. For the naturalist, there is no transcendent source that defines right and wrong. For the strict naturalist, morality is purely a human creation formulated to make society livable. Morality has a place in that it enables individuals and societies to survive and even flourish. According to the biologist E. O. Wilson, "Ethical codes have arisen by evolution through the interplay of biology and culture."[10] Morality, according to the naturalist, is the result of purely material processes and ultimately has a biological basis.

The view that human beings are nothing more than the sum of their physical parts and properties is a tenet of physicalism. The naturalist philosopher Paul M. Churchland puts it this way:

> The important point about the standard evolutionary story is that the human species and all of its features are the wholly physical outcome of a purely physical process. If this is the correct account of our origins, then there seems neither need, nor room, to fit any nonphysical substances or properties into our theoretical account of ourselves. We are creatures of matter. And we should learn to live with that fact.[11]

Finally, for the naturalist, history has no direction, goal, or purpose. It is simply the continuation of events begun randomly in the past. It is no more than the record of one event following another. Indeed, for the naturalist, there is no teleology at all, and thus even human actions are not literally done for a purpose or end. However, within the scientific community, there is often high optimism about scientific and technological progress. Nowhere is this more evident than in the biotechnology community. An incurable optimism about where the biotechnology revolution is going can lead the members of this community to minimize potential problems. But this optimism is not related to any overall view of where history is going.

Worldview and Genetic Technologies: The Human Genome Project

One of the clearest examples of the materialist worldview at work in biotechnology involves the genetic revolution. The explosion of genetic information and the contribution of genes, or groups of genes, to conditions of personality have suggested anew that the essence of human beings can be reduced to their genetic material; that the core of a human person is now found in his or her genome.

This view of a human person is not unusual in the community of scientists working on the Human Genome Project (HGP), which was established in 1989 and is now part of the federal National Institutes of

Health—the National Human Genome Research Institute. The draft of the entire human genome was completed in the spring of 2003, and analysis and interpretation have begun. We cannot survey every scientist working on the project, and there will undoubtedly be exceptions to this general rule. But numerous genome researchers and some of the project's leaders have written publicly on their role in the project and how they view human nature as a result of their work. Some of the most prominent leaders have backed away from earlier and more radical statements about genetics and the human person, but there is no reason to believe that their basic materialist assumptions about human personhood have changed.[12] The historian of science Dorothy Nelkin and the sociologist Susan M. Lindee summarize the views of the genome research community when they suggest that "just as the Christian soul has provided an archetypal concept through which to understand the person and the continuity of self, so DNA appears in popular culture as a soul-like entity, a holy and immortal relic, a forbidden territory."[13]

For example, the Harvard molecular biologist Walter Gilbert has called the genome the "holy grail" of human identity and has suggested that understanding the human genome provides the ultimate answer to the commandment "Know thyself."[14] He suggests that as a result of the genome project, our philosophical views of a human person will change. Clearly reflecting a materialist view of the world, which implies that human personhood is reducible to genetics, he argues that "to recognize that we are determined, in a certain sense, by a finite collection of information that is knowable will change our view of ourselves."[15] Gilbert paints the scenario in which, at the completion of the genome project, a person could store all their genetic information on a single compact disc (CD). He concludes that "three billion bases of sequence can be put on a single compact disk, and one will be able to pull a CD out of one's pocket and say, 'Here is a human being, it's me.'" Gilbert clearly sees the essence of a human person in the genetic material to be unearthed by the genome project, such that when the project is finished, society will have "closed an intellectual frontier, with which we will have to come to terms." He puts this even more clearly when he suggests that "knowing the complete human genome we will know what it means to be human."[16]

James Watson and Francis Crick, the codiscoverers of the structure of the DNA molecule in 1953, have expressed similar views. Crick has suggested that "you, your joys and your sorrows, your memories and your ambitions, your sense of personal identity and free will, are in fact no more than the [genetically determined] behavior of a vast assembly of nerve cells and their associated molecules."[17] That is, in his view, everything that makes a human being a person is determined by biological processes that are driven by one's genetic structure. His longtime colleague Watson has argued that the HGP has as its ultimate aim the search for "ultimate answers to the chemical underpinnings of human existence."[18] That statement, taken by itself and referring to the chemical underpinnings of human existence, is not necessarily indicative of a kind of genetic reductionism, but his elaborations elsewhere are. "We used to think our fate was in the stars," he has asserted. "Now we know, in large measure, our fate is in our genes."[19]

More subtle ways of expressing this assumption about one's humanity being reducible to one's genetic material appear in the way the genome researchers talk about the project. Their metaphors and figures of speech suggest a reductionist view of the human person.[20] For example, molecular biologists routinely describe the life processes they investigate in terms of information, messages, and codes. DNA is commonly referred to as a program, a text, or more frequently a code by which the machinery of the body operates. Cells are commonly referred to as chemical factories that are operated by the computer program of DNA. Organisms become information systems that are analogous to computer systems in the storage of and activity based on the information available to them from the master molecule, DNA. Cells, then, become machines that run according to the genetic instructions given to them, a view that is echoed strongly by more futuristic biotechnology advocates.

The choice of these figures of speech reflects a materialist view of the world, with genetic reductionism as the best way to capture that view. The sociobiologist Howard L. Kaye has suggested that molecular biologists' perception of life in terms analogous to a computer represents a worldview that sees life in materialist terms. He argues that "the reduction of all of biology, all of the behavioral characteristics and fundamentals of living things to molecular mechanisms of life betrays a metaphysical ambition to demonstrate that organisms really are machines, and that all of life can be accounted for in this way."[21] Similarly, in a speech to the International Congress of Genetics in 1988, the late geneticist and biophysicist Robert H. Haynes argued that the advent of genetic technologies proved his materialist view of the world. As the president of one of the largest of these conferences, he put it like this:

> For three thousand years at least, a majority of people have considered that human beings were special, were magic. It's the Judeo-Christian view of man. What the ability to manipulate genes should indicate to people is *the very deep extent to which we are biological machines* [emphasis added]. The traditional view is built on the foundation that life is sacred. Well, not anymore. It's no longer possible to live by the idea that there is something special, unique, even sacred about living organisms.[22]

As noted above, some who are involved in the HGP do not share such a view of the world, and others are interested only in the clinical application of its findings.[23] But it is clear that this view of the world is widely shared in the genetics community and has an influence on the work being done.

Worldview and Biotechnology: Human Beings as Machines

The materialist view of the world is equally evident in the more futuristic predictions of the advocates of biotechnology. Naturalistic evolution is presupposed in virtually all this literature, and advocates of biotechnology speak of the *transhuman* person as the next step in the evolutionary process, on the way to a new species, the *posthuman* person. For example, the futurist Kevin Kelly suggests that there is a parallel between Darwinism and what he calls "the neo-biological civilization," a parallel between natural evolution and what he calls *artificial evolution*."[24] As he puts it, "Natural evolution insists that we are apes; artificial evolution insists that we are machines with an attitude."[25] As we hope will become clear, given this fundamental assumption, it is not surprising that such futurists, or transhumanists, see human beings in mechanistic terms.

Kelly makes this explicit when he suggests that the movement to neobiological civilization will have social consequences. "The greatest social consequence of neo-biological civilization," says Kelly, "will be the grudging acceptance by humans that humans are the random ancestors of machines, and *that as machines* [emphasis added] we can be engineered ourselves."[26] Viewing human persons as machines that can be programmed and improved on is the assumption on which Ray Kurzweil builds his book *The Age of Spiritual Machines: When Computers Exceed Human Intelligence*.[27] Kurzweil argues that computers will at some point in the future exceed the capabilities of human intelligence, as he argues is being done in simple ways such as computational skills and memory. He assumes an evolutionary framework and finds it remarkable that this "apparent tautology went unnoticed until a couple of centuries ago." He makes this plain when he calls the universe "a closed system (not subject to outside influence, *since there is nothing outside the universe*)" (emphasis added).[28] Kelly echoes this when he defines life as "a nonspiritual, almost mathematical property that can emerge from a networklike arrangement of matter."[29] Katharine Hayles suggests something similar when she defines the posthuman person as "amalgam, a collection of heterogeneous components, *a material-informational entity* [emphasis added] whose boundaries undergo continuous construction and reconstruction."[30] That is, a posthuman person is nothing more than a collection of material parts and information, which Hayles suggests can eventually be seamlessly integrated with machines. She argues that "there are no essential differences or absolute demarcations between bodily existence and computer simulation, cybernetic mechanism and biological organism, robot teleology and human goals."[31] Clearly, this view of a person is possible only on the assumptions of the scientific materialist, which reduce human persons to their biological parts and biotechnological enhancements.

This reductionism is reinforced by Damien Broderick in his discussion of the mind-body problem and the phenomenon of consciousness when he argues for equating the mind with the brain on the basis of the dominant scientific materialism of the cognitive neurosciences:

> The fleshy body is not a prison for some impalpable mind-stuff. That is the error of dualism; philosophy and science alike have taken centuries to escape it. Mind is flesh, *we know of no other kind* [emphasis added].[32]

Broderick recognizes that reducing consciousness to material properties is the subject of significant philosophical debate today yet accepts the system that he calls "the unfinished materialist account of cognitive science." He contrasts what he calls "ancient dogmas" from both Eastern and Western religious traditions supporting an immaterial aspect in a person with what he calls *transhumanists*—"materialists who maintain that mind is indeed nothing other than the sublimely complex workings of the physical brain and its bodily extensions." He asks: "If that is what we are, what is to prevent us from copying—mapping—our neurological complexity into some more durable, swifter material substrate?"[33] This process he calls *uploading*, a concept addressed by a variety of the futurist advocates of biotechnology, and it too presumes a materialist view of the world. He does exactly that when he presupposes that "mind and passion and soul are indeed the body, that whirling composite of matter and force and energy, in motion in the world."[34] Here he clearly reduces what are commonly taken to be immaterial components of a person to their physical properties and argues that that is all they are.

One technology envisioned by many of the futurist advocates of biotechnology is the process of uploading the information contained in one's brain into another body. Kurzweil and others suggest that at some point it will be possible not only to have neural enhancements but also to scan the brain with its entire neural system and transfer it to a computer or to another body, a process they term *instantiation*. Although Kurzweil argues that identity is not a function of the parts but rather "the pattern of matter and energy we represent," identity is still ultimately reducible to the materia.[35] He envisions what would happen if a person's neural material were scanned and transferred, then subsequently destroyed after it had been instantiated in a new body or computer. This, he suggests, might be killing the person or having that person commit suicide, which implies that the essence of the person is subsumed in the brain. He clarifies this equation of the person with the brain when he argues that when the neural material is instantiated in another medium, a "newly emergent person" will appear. In his work, he links spiritual experience with the brain and suggests a neurological basis for spiritual experience, insisting that the next generation of machines will "connect with their spiritual dimension."[36]

In his most recent volume, the massive 650-page *The Singularity Is Near: When Humans Transcend Biology*, Kurzweil prophesies that the "singularity" may be upon us by 2045. The Singularity is a future period that "will represent the culmination of the merger of our biological thinking and existence with our technology, resulting in a world that is still human but that transcends our biological roots." In this world, maintains Kurzweil, "There will be no distinction . . . between human and machine or between physical and virtual reality."[37]

Limitations and Contradictions of Philosophical Naturalism

It would be difficult to imagine a worldview more at odds with Christian theism than philosophical naturalism. In terms of Christian metaphysics, God is the ultimate reality, and there is a realm of reality that is not material. Epistemologically, what can be empirically verified clearly does count for knowledge but knowledge is not limited to that realm. God reveals Himself and His truth both through general revelation in His world and through special revelation in His Word. Those aspects of revelation also count for real knowledge and cannot be discounted or marginalized as "belief." From a Christian perspective, human beings are made in God's image, the result of the intimate special creation by God, not random forces apart from divine intention and intelligence. Human beings have special significance by virtue of being in God's image, and life has purpose and meaning that revolves around bringing honor to and knowing God. Human beings have genuine libertarian freedom that undergirds the notion of moral responsibility and criminal justice. Morality comes ultimately from the character of a transcendent God, who issues commands consistent with His character. Right and wrong are not fundamentally human creations and are not fluid and changing with changing times and cultures. Finally, history has a definite direction because God will bring history to its culmination with the return of Christ.

Of course, arguing that naturalism is inconsistent with Christian theism is unlikely to carry much weight with the naturalist. But there are other aspects of the naturalist view of the world that deserve comment. Our observations are consistent with Christian theism but not dependent on it. In other words, one does not have to accept Christian theism as a worldview to have philosophical reservations about naturalism.

A first observation is that the naturalist's fundamental thesis cannot meet the test of its own criteria. The statements "All reality is material" and "All knowledge is that which is empirically verifiable" are themselves not empirically verifiable. Thus, these statements are self-refuting. They are not scientific statements but rather philosophical statements about science. Yet they are treated as unassailable facts instead of the philosophical commitments that they are. For the materialist to discount the theist's worldview because it is a faith commitment is disingenuous at best and hypocritical at worst.

One can see evidence of this flaw in scientific reasoning in the discussion in genetics. The scientists who are reducing the notion of personhood to one's genetics are making a fundamental methodological error in expecting biological data to provide the foundations for metaphysical questions. Biology and biochemistry provide the raw material for biological and biochemical questions. They cannot decisively answer what are ultimately metaphysical questions. When molecular biologists make such metaphysical claims about the nature of personhood, they are no longer practicing molecular biology. They have become metaphysicians and are inappropriately extending their expertise from biology to metaphysics. One simply cannot extrapolate metaphysical conclusions from biological and biochemical data without leaving one's scientific field. The scientists who are arguing for some sort of genetic reductionism are imposing their view of the world on their work and drawing metaphysical conclusions that simply cannot be drawn from their field of expertise.

A second concern about naturalism arises in the moral and social implications of such a worldview. The way in which people are held morally and criminally responsible for their actions strongly suggests that they possess free will. This is a necessary presupposition for any concept of moral responsibility to have meaning. That is, people make genuinely free choices, for which they deserve praise or blame. Society's notion of criminal justice is premised on such a view of a human person.

Naturalism implies determinism, and such a view undermines important notions of free will and attendant moral responsibility. If all events are caused by prior events (known as *event causation*), then there is no room in the system for free agency in which a person is a first or unmoved mover acting freely. Of course, one may be influenced by desires, past events, and so forth, but that is very different from insisting that one's actions were caused by those things. The concept of character is premised on freedom to make one's choices, which thereby contribute to shaping identity. To take determinism seriously, as would seem to be implied by a naturalist view of a human person, would undermine critical moral and legal notions of personal and criminal responsibility. Some have tried to argue that genuine freedom can emerge in a strictly physical universe, but these claims are unpersuasive.[38] The only notion of freedom existing for the naturalist is unpredictability at the quantum level, but that at best is an unpredictable determination.

A further difficulty for the naturalist in the area of moral responsibility comes from the failure of physicalism to account for personal identity. If a human being is nothing more than the collection of physical parts and properties, then nothing like the soul exists, and no notions of this concept make any sense given a physicalist view of a human being. More important, for the physicalist there is nothing to ground personal identity through time and change. Some physicalists try to deny this charge, arguing that even if a human is merely physical, there is still a robust doctrine of personal identity available. But it is hard to see how this could be the case because an aggregate of parts cannot remain the same through loss of old parts and gain of new ones. There is nothing to ground sameness. That is why many physicalists reject a robust view of personal identity. For many physicalists, when an entity—including a person construed materialistically—loses some parts and gains new ones, it literally becomes a different entity. But a person maintains his or her personal identity even though body parts are lost and even though a person will get a new set of cells approximately every seven years. To be consistent, the physicalist would have to maintain that this person is a different person. Yet our notion of moral responsibility and criminal justice is built on the concept of enduring personal identity through change. For example, if a person commits a crime and is not brought to trial until several months to a year later, under a physicalist view of a human being, one could insist that he or she is a different person than the one who committed the crime. Clearly such a view is absurd, and a worldview that recognizes an immaterial aspect to a human being, such as a soul, and grounds personal identity through time and change offers a more plausible explanation.

A further point of tension in ethics arises in the grounding for the critical notion of human dignity. In an evolutionary view of the world holding that human beings are the products of matter, energy, and physical processes, with no such thing as a human nature—not to mention anything like the image of God—human dignity has nothing on which to ultimately ground it. If human beings are the product of blind, random evolutionary forces, then there is no fundamental reason to treat human beings with any respect or dignity at all. It is not an accident that the notion of human dignity arose in settings informed by a view of a human being as having an essence that grounds one's identity. More specifically, the modern notions of human dignity arose out of a commitment to human beings being the product of a Creator, made in His image, and today, as naturalism gains dominance, human dignity is clearly a holdover from a Judeo-Christian heritage that has long been abandoned. Contrast that with the record of the majority of the most heinous human rights abuses in the twentieth century occurring in avowedly atheistic regimes such as communist nations. One cannot have a human nature in a physicalist view of a human being. All one has is biology, and biology cannot ground concepts such as human dignity.

But there is a more fundamental problem: The naturalist cannot account for moral properties and value. The philosopher J. L. Mackie concedes that moral properties are problematic for naturalism when he says that "moral properties constitute so odd a cluster of properties and relations that they are most unlikely to have arisen in the ordinary course of events without an all-powerful god to create them."[39] The naturalist cannot make the statement such as "Compassion is a virtue" and have it be a moral norm with truth value, because this statement cannot be empirically tested and does not deal with the kinds of properties that could arise out of a strictly physical universe. The naturalist can offer "Compassion is a virtue" as a statement of preference ("I like compassion"), as moral cheerleading ("Hooray for compassion!"), or as a factual statement, such as "Compassion promotes society's survival," but not moral norms with truth value. This leaves the naturalist with either a sense of nihilism about ethics or an acceptance of making ethics subjective to one's personal choice. The latter option takes the idea of the good life and good person to be a matter of preference, as long as they "do no harm" (ironically, the moral absolute with clear truth value for the naturalist). But such subjectivity leads to wildly counterintuitive conclusions that there is no real difference between a life lived in utter narcissism and one of altruism and compassion. Ultimately, naturalism provides no adequate answer to the critical question "Why be moral?"

The Worldview of Environmentalist Biocentrism

Far removed from the materialists with their uncritical enthusiasm for biotechnology stand the radical environmentalists, who oppose most if not all biotechnological interventions. This group of thinkers and activists has focused most of their efforts on genetic technologies, applied to animals, foods, and human beings. They argue that such technologies are incompatible with sustainable growth, will not feed the world or save the environment, pose unacceptable risks to the environment, and allow large corporations to exercise monopolistic control over food supplies and human health.[40] Many opponents of biotechnology also oppose the forces of globalization, which they see as inextricably woven together with the advances of biotechnology. Some of these opponents point out the naturalism that dominates the scientific-economic interests of proponents of biotechnology.[41] But they too have their distinctive view of the world undergirding their environmental concerns about biotechnology.

Biocentrism, also known as deep ecology, represents a view of the natural world in which the environment and certain species of animals (and also, for some, plants) have intrinsic value as opposed to instrumental value. Deep ecologists hold that the environment can and should be protected for its own sake, not for its benefits to human beings. Many environmentalists and religious groups with environmental concerns have adopted this view and have given it strongly spiritual overtones. Creation spirituality, the Mother Earth movement, and treating the planet as a sacred thing to be worshiped are all outgrowths of biocentrism. Some suggest that these biocentric views are actually a form of ancient paganism, in which the creation was revered and worshiped.[42] In more extreme forms, ecoterrorism is sometimes referred to as a "form of worship," with many movements supporting an emphasis on "bonding" with the earth. In this view, ultimate reality is the natural world, which is endowed with a sacred quality that human beings are responsible for respecting, thus placing significant limits on how the environment can be used for human benefits. People who hold to a biocentric view of the world object to most biotechnology, including genetically modifying foods and animals.

Biocentrism also holds that there are few if any moral differences between human beings and animals—that animals, and some plants perhaps, are entitled to the same moral considerations and respect owed to human beings. Further, this worldview tends to see history as cyclical rather than linear, which may reflect the influence of Eastern thought. In sum, the ultimate purpose of the environment is not necessarily or even primarily to benefit human beings, not to mention global corporations and their profits.

Biocentrism, Biotechnology, and the Purpose of Life

The biocentrist view comes out most clearly in considering the purpose for life on earth. For example, in their critique of genetically altered foods, under the general heading of a clash of worldviews, Marc Lappe and Britt Bailey take on the corporate worldview (which others call *androcentrism*):

> Some religious figures (who reflect biocentrism with a religious undercurrent) declare the belief that nature is intended for man's use engenders a patriarchy of control that misconstrues *the purpose of life on earth* [emphasis added]. Others decry the attitude that casts all living things as intrinsically designed for our exclusive use (androcentrism). The truth,

> of course, is that the natural world was never intended or designed for exclusive use, nor shaped to fit our nutritional needs.[43]

This reasoning follows from believing that the environment, or at least parts of it, has moral status because of its intrinsic value. For example, in their critique of genetic modification of foods and farm animals, Sheldon Krimsky and Roger Wrubel identify two different types of biocentrism—*biocentric individualism* and *biocentric holism*. The former assigns moral value to individual nonhuman things within the environment. They state:

> [The] biocentric designation indicates that some classes of nonhuman entities have intrinsic moral worth. It extends the notion of rights and moral obligations associated with the theory of persons to other forms of life. All life forms with some set of designated characteristics (sentience, cognitive capacities, teleological centers of activity, and response to environmental stimuli) have moral status under biocentric individualism.[44]

To extend this further, biocentric holism derives moral value for nonhuman entities from the concept of a species, with such moral value "egalitarian across species." The difference with the former is that biocentric holism allows for interests of species to take precedence over individual organisms.[45] Protecting species diversity or species integrity would be weighted more heavily than the interests of individual entities in the environment.

The interest in biodiversity is often driven by a biocentric view of the environment. Vandana Shiva argues for protecting biodiversity on the basis of the rights of the environment. She suggests that "the conservation of biodiversity, *at its most fundamental level* [emphasis added] is the ethical recognition that other species and cultures have rights, that they do not merely derive value from economic exploitation by a few privileged humans."[46] That is, nonhuman beings and entities have rights that ought to be protected because they have value independent of how human beings use and benefit from them.

Still others, such as Martin Teitel and Kimberly A. Wilson, refer to biocentrism as an "ecological set of values."[47] These include sustainable agriculture, harmony between humans and nonhumans, avoiding the ends of efficiency and production, balance, integration, intrinsic function of the environment, and humility toward it. Teitel and Wilson contrast their ecological worldview with those of "engineering" and "religion," reflecting the competing worldviews of materialism, mentioned earlier, and Christian theism, discussed ahead.[48] They suggest a more fundamental part of their worldview as a philosophical division "between those who see themselves as part of nature, and those who see themselves as standing outside of nature, in a special category, not subject to the rules or processes that seem to govern life in our world."[49] Those who see themselves as part of nature, with no special human moral status that sets human beings apart, see the world from a biocentric perspective, with intrinsic value and concomitant rights that must be protected. The biology professor David Ehrenfeld affirms this when addressing the value of species, that they simply have value independent of the uses and properties they have. He puts it like this:

> Value is an intrinsic part of diversity; it does not depend on the properties of the species in question, the uses to which particular species may or may not be put, or their alleged role in the balance of global ecosystems. For biological diversity, value is perhaps nothing more, and certainly nothing less.[50]

Implications of Environmental Biocentrism for Human Dominion

Environmentalist biocentrism, like scientific materialism, is at odds with the worldview of Christian theism. A Christian environmental ethic combines the biblical themes of human dominion over creation and responsible stewardship to take care of it. The creation has value because it is God's creation, promoting what some call a "theocentric" view of nature.[51] To be sure, God reveals Himself in the creation (Ps. 19:1) and the Earth does belong to the Lord (Ps. 24:1). But the Earth is not the Lord. Nowhere does the Bible or any Judeo-Christian ethic that is consistent with the Bible equate worship of the creation with worship of the Creator. In fact, one of the purposes of the Genesis account of creation was to distance Old Testament theology from the

Caananite religions of the Middle East, most of which worshiped the creation or parts of it. The creation account in the Bible is very clear that God stands over and above the creation. He is not to be identified with creation, nor is creation to be worshiped instead of Him. We honor God when we properly care for His creation, exercising our role as stewards over it. But we also honor God when we exercise our dominion over creation, developing it and harnessing it for the benefit of humanity. There is no reason why a proper view of responsible dominion over the environment, balanced by the responsibility to be stewards of creation, cannot produce a genuine environmental concern.

A second concern with biocentrism is that it appears to lead to the notion of trees, plants, and animals having parallel rights with human beings. The ethicist W. Michael Hoffman suggests that nonhuman living things are also integral parts of the ecosystem that have intrinsic value.[52] Human beings with rights do not stand above animals and plants, lacking such rights. Rather, all are part of a more holistic system in which all things are valued equally. Hoffman and others, such as the philosopher Peter Singer, would charge that a view that places greater intrinsic value on human beings is "speciesism."[53] Biocentrism would seem to lead to the idea that animals and trees have rights that should be protected. We would not want to suggest that animals, for example, have no interests that are not worthy of protection. We would hold that cruelty to animals is immoral, but we would stop short of insisting that animals have rights. There is a good deal of debate over animal rights that is beyond the scope of this discussion, but at the least we would suggest that animals and plants are not rights bearers but that animals may have some interests that should protect them from cruelty. The problem with the way biocentrists view plants and animals is that it presents a system that is very difficult to live with consistently. To be fair, Hoffman does hold that with clear criteria human interests can take precedence over the environment, but it is not clear what exactly those important criteria are. However, once one admits that animals and trees have rights, then it would seem to be difficult to draw the necessary lines that would justify promoting human interests ahead of those of the rest of the ecosystem.

The Worldview of Christian Theism

The theistic worldview of the Bible stands in sharp contrast to the culturally dominant worldview of philosophical naturalism. This chapter introduces the main contours of Judeo-Christian theism, with the following chapters fleshing out its application to various aspects of biotechnology. Space does not allow for a detailed philosophical defense of each of these components.[54] Although this theistic worldview is fundamentally Christian, it has a good deal in common with Judaism and shares some common elements with other theistic religions such as Islam.

The Creator and His Creation: Ultimate Reality and Materiality

For the Christian, ultimate reality is fundamentally spiritual and immaterial. God is the ultimate reality, and the material world is the product of His special creation, not the product of random forces of evolution apart from an intelligent designer. The material world is not eternal, nor is it divine. The Genesis account of creation makes it very clear that there is a critical dichotomy between creation and Creator, and it is the Creator, not the creation, who is to be worshiped. The material world is good and has value because it is God's creation (Gen. 1:31; 1 Tim. 4:1–4) but also because of the Incarnation. In the Incarnation, God inhabited the material world in human flesh (Col. 1:15; John 1:14). The Incarnation was a decisive argument against many of the early church's heresies that sought to elevate the soul at the expense of the body. The material world has value, although not as ultimate reality.

An Epistemology of Metaphysical Realism

The Bible is clear that the material world can teach human beings some things about the reality and nature of God (Ps. 19:1–7) and that the structure of the world reveals the wisdom of God (Prov. 8:22–31). Theologically, this is known as *general revelation*. In addition, the Bible teaches that the material world is real and not an illusion. Known as *metaphysical realism*, this view of reality maintains that the material world really does exist, that it can be known, and that what human beings can know about the world actually corresponds to what exists. Reality is not fundamentally a social construction, although the ways in which we describe reality are influenced by a variety of factors.

Although the Bible maintains that there are a variety of ways in which human beings can know truth, foundational to a Christian worldview is the notion that truth does exist in a real and objective sense. The Bible maintains that God's Word is truth (Ps. 119:160) and that Jesus is the bodily expression of God's truth (John 14:6). These suggest that knowledge cannot be reduced to what is scientifically or empirically verifiable, nor is there as sharp a distinction between knowledge and belief as naturalists would presume. To be sure, what is empirically verifiable clearly counts for knowledge. Through general revelation, God reveals His truth. But it is clear in a Christian view of the world that knowledge goes well beyond what the senses and science can verify. God reveals His truth in His Word as well as in His world (Ps. 19), although in both arenas of revelation human beings' ability to apprehend God's truth has been affected by sin. That is, the lenses through which human beings see both God's Word and God's world are clouded, although not to the degree that human beings are incapable of knowing God's truth.[55]

In terms of epistemology, moral values have objective truth value and cannot be relegated to subjective opinion or belief. For example, the moral propositions in the Decalogue are taken to be truth and have been made known by means of special revelation. In the Old Testament, the Mosaic Law was God's truth—much more than matters of subjective belief only—and the law clearly counted for real knowledge. Similarly, the Sermon on the Mount cannot be reduced to mere belief but rather contains moral and theological knowledge regarded by Jesus as truth. Scripture offers categories for moral truth that cannot be reduced to subjective opinion or preference.

General Revelation and the Dominion Mandate

The concept of general revelation lays some of the groundwork for our view of technology. Theologically, it is not an accident that human beings have made the discoveries of science and their application in terms of technology. The Bible is clear that God has embedded His wisdom in the world (Prov. 8:22–31) and through general revelation has given human beings the tools to unlock and apply His wisdom. This is part of the dominion mandate given to the human race (Gen. 1:28). In the prelapsarian Garden, human beings existed without the corruption and decay, toil and trouble that resulted from God's judgment upon the earth for human sin (Gen. 3:15–19). To be sure, in the Garden, Adam and Eve were to "keep" and "till" the earth; but the labor was not "laborious." After the fall into sin, things became more complicated. Work became toil*some* (Gen. 3:17). The ground was filled with thorn and thistles (Gen. 3:18) and, thus, the extension of dominion over the earth came to include efforts to alleviate the effects of sin. For example, medical technology is God's gift to human beings to alleviate disease, clearly one of the effects of the entrance of sin into the world. Many biotechnological innovations also fit under this heading.

A Christian Anthropology: The Image of God, Human Dignity, and History

Critical to the revolution in biotechnology is the view of a human being—our anthropology. Human beings are a body–soul unity having both material and immaterial aspects. The Bible clearly teaches that human beings are more than simply their bodies, more than a collection of parts and properties. The Bible is clear that human beings have souls, which animate the body and provide the capacities for a relationship with God. This stands in sharp contrast to the physicalism that pervades the sciences today. Further, human beings have intrinsic value by virtue of being made in God's image (Gen. 1:26–27), which distinguishes human beings from animals. The image of God, a doctrine often referred to in Latin as *imago dei*, is fundamental to the notion of human nature, which grounds personal identity through time and change. In addition, the image of God grounds the sanctity of innocent human life (Gen. 9:6) and the principle of respect for human dignity and equality. Human beings have essential dignity by virtue of being made in God's image. This is the ultimate basis for human rights and the reason that society ought to be concerned about the various ways biotechnology threatens to undermine human dignity.[56]

Of course, this view of the image of God and human dignity also has room for appreciation and encouragement of biotechnologies that uphold human dignity and work toward alleviating the effects of the entrance of sin into the world. Because of the reality of sin, the image of God has been tarnished, and death and decay are part of the realities of life on earth. Thus there is a need for technological innovations that can both relieve human suffering and at the same time safeguard fundamental human dignity.

The notion of human dignity is the ethical principle often invoked to challenge the seemingly inexorable march of biotechnology. In a Christian worldview, morality is objective and universal and has a transcendent source. It is not fundamentally a human creation. That is, morality is discovered or discerned, not created. God has revealed universal moral virtues and principles, both in the Bible and in His world (the latter known as *natural law*). The Bible gives both specific commands and general principles that guide action and can be applied to the various biotechnological options being considered. Both the commands are principles ultimately rooted in the virtues that are reflected in God's character, which is the ultimate source of morality.

In a Christian worldview, death is not the end of life. Again, this is a vivid contrast to the dominant naturalism of the culture, which holds that there is nothing beyond one's physical life. For the naturalist, death is the end, and there is nothing apart from earthly life for which a person should or could live. In scripture, however, there is an eternal hope that serves as motivation for the believer to live faithfully in line with God's design. In fact, there is as much hope for one's body as there is for one's soul, for the promise of the resurrection at Christ's return involves a transformation of the body (1 Cor. 15:35–58). The Bible indicates that between one's death and the Lord's return, the believer is present with God in an intermediate state and is rejoined with his or her glorified body at the second advent of Christ (2 Cor. 5:1–10). Because of the hope that exists for the body, there is a basis for taking care of the body during life on earth (1 Cor. 6:19–20) and for applying biotechnology within limits to alleviate the effects of sin. This contrasts with some of the ancient heresies that denigrated the body and upheld the soul as that which really mattered to God.

A final component of a Christian worldview is the view of history. For the Christian, history has a direction and an end, which will be completed at Christ's return. Again, this is quite a contrast to the view of the naturalist, who insists that history is going nowhere and that events are unrelated to any order or direction. Some biotechnology advocates even speak of using biotechnology to move human beings to the next stage of evolution, thereby enabling human beings to write their own history. In a Christian worldview, history has a goal and a direction (Eph. 1:9–10). It is a symphony of which God is the composer and the conductor (Eccles. 3:1–9). Moreover, there will be a judgment in which human beings are held accountable for how they have behaved toward one another and how they have respected human life (1 Cor. 3:11–15; Heb. 9:27).

Conclusion

It is critical to pay attention to the unexamined worldview assumptions in discussing biotechnology, particularly given the naturalism that pervades the scientific and biotechnology communities. The debate is not only over the specific biotechnologies that are being considered. There needs to be a broader discussion about the underlying views of the world held by both advocates and opponents of various biotechnology scenarios. Ideas do matter, and they shape the ways the culture thinks about biotechnology and its applications. Crucial to that discussion are the competing views of a human person, his or her value, and the nature of the environment. The philosophical naturalism underlying much of the scientific and biotechnology community contributes to its general uncritical enthusiasm for biotechnology in the same way the biocentrism of the environmentalist contributes to the general negativity toward biotechnology. The measured enthusiasm of the Christian theist is grounded in both its respect for human dignity that comes from human beings being made in the image of God and its view of responsible human dominion over the environment.

CHAPTER FOUR

Biotechnology and Human Dignity

> Recognition of the inherent dignity and of the equal and inalienable rights of all members of the human family is the foundation of freedom, justice, and peace in the world.
>
> —United Nations Universal Declaration of Human Rights

FEW terms or ideas are more central to bioethics—or less clearly defined—than *human dignity*.[1] People invoke it to support almost anything. So it is a standard to which most people are quite receptive. But understanding how people use it, and how a Christian perspective on it can help clarify its meaning and implications, constitutes one of the greatest challenges and opportunities in bioethics today. Particularly in the face of emerging biotechnologies, we need a clear anthropology—a clear understanding of who human beings are and, by implication, how they may and may not act toward one another and the world.

Clarifying what human dignity is all about is a wonderful vehicle for developing such understanding, in part because the concept is so important to so many—nonreligious and religious people alike. Respect for human dignity is an ethical mandate to which both sides of many bioethical debates appeal. For example, the state of Oregon legalized physician-assisted suicide by passing the Death with Dignity Act, but opponents claimed that legalizing the practice would instead undermine the dignity of elderly, disabled, and dying patients. Similarly, opposing claims are made in response to assertions that respect for the dignity of these same patients demands pursuing cures by producing embryos for embryonic stem cell research via cloning: Producing human beings in embryonic form and destroying them for the benefit of others is an affront to human dignity.

The term "human dignity" is also surfacing more frequently in significant bioethics and other public documents such as national constitutions and international bioethics agreements. It has played a role in the constitutions of a politically diverse array of countries, including Afghanistan, Brazil, Canada, Costa Rica, the former Federal Republic of Germany, Greece, Guatemala, Ireland, Italy, Nicaragua, Peru, Portugal, South Korea, Spain, Sweden, and Turkey. In some of these countries, such as Germany, the role of human dignity is quite substantial. Affirming that "the dignity of the human being is inviolable," the German Constitution recognizes various human rights that the law must respect.[2] Even in countries where the term has not been influential in constitutional language, it has come to play an important role. For example, the U.S. Supreme Court has employed the term in its deliberations over the meaning of the First, Fourth, Fifth, Sixth, Eighth, and Fourteenth amendments to the Constitution.

International documents that are relevant to issues in bioethics similarly have affirmed the critical importance of human dignity. The United Nations, whose very charter celebrates the "inherent dignity" of "all members of the human family," issued a Universal Declaration of Human Rights in 1948 with a preamble containing the same language.[3] The Universal Declaration's Article 1 more specifically affirms that "all human beings" are born "equal in dignity." Two other documents—the International Covenant on Economic, Social and Cultural Rights and the International Covenant on Civil and Political Rights—joined this document in 1966 to constitute the so-called International Bill of Rights. All three documents ground the various rights of all human beings in their human dignity. In line with this outlook, the Council of Europe's 1996 Convention on Human Rights and Biomedicine was designed explicitly to "protect the dignity" of "all human beings."[4]

These documents reflect the primary sense in which human dignity is invoked today: as an attribute of all human beings equally that establishes their great significance or worth. The word *dignity* comes from the Latin words *dignitas* ("worth") and *dignus* ("worthy"), suggesting that dignity points to some standard by which people should be viewed and treated. Though the standard usually has an egalitarian bent to it today, in the world of ancient Greece and Rome the standard more commonly was attached to inegalitarian traits, such as physical prowess and intellectual wisdom—as exemplified in such figures as Hercules and Socrates. People differed in dignity according to the degree to which they manifested the relevant traits, and the honor due them varied accordingly. This sense of dignity persists today when one speaks of "dignitaries" who warrant special honor, or of behaviors or conditions that are dignified or undignified. Dignity in this sense can increase or decrease, can be attained or lost.[5]

Dignity can refer to something that is variable in other ways as well. There is a difference between having dignity, on the one hand, and having an awareness of dignity or being treated with dignity, on the other. Someone may not be aware of having dignity, although possessing it nevertheless; someone may not treat people in a particular group as having dignity, though they may indeed possess it. Such variability, however, should not be confused with the contemporary concept of dignity, which is beyond the perceptions or actions of particular individuals and is rooted in what all human beings have in common. Such is the concept typically operative when *human* dignity is invoked as the basis for how human beings should be viewed or treated.

Respect for human dignity is connected to a virtue as well as to an ethical standard. A virtue-oriented approach to human dignity may take different forms. For example, exhibiting human dignity (usually referred to simply as "dignity") can be a virtue in a way reminiscent of the above-mentioned notion of "dignified behavior." To say that certain people exhibit dignity or are dignified can be a way of commending their courageous attitudes or honorable actions in the face of adversity.[6] Conversely, the virtue of human dignity may refer to a person's capacity to recognize and live in accordance with a particular standard of human dignity. This form of the virtue serves as a reminder of how important it is that respect for human dignity be lived out in practice rather than exist merely as an abstract concept. Nevertheless, exercising such a virtue does require specifying what human dignity is.

People most commonly view human dignity in one of two basic ways. Some see it as rooted in particular characteristics of human beings; others view it as attached to being human per se, regardless of any particular characteristics they may have. Both understandings are examined in this chapter, followed by a survey of some of the bioethical implications of such views. First, however, it is helpful to clarify the significance and meaning of the concept by noting arenas in which it has been challenged.

Challenges to Human Dignity

In the twentieth century, perhaps the most widely decried denial of human dignity took place under the fascist regime in Germany—which accounts for the strong emphasis on dignity in the German Constitution and international documents noted above. Millions of people were forced to be subjects of experimentation against their will or were tortured or killed. As a result, the importance of human freedom and bodily integrity became much clearer, and the danger of compromising them in the interests of the larger society became widely evident.

In fact, a tension necessarily exists between the idea of human dignity and ethical outlooks such as utilitarianism that (at least in their more popular and influential forms) affirm human dignity only to the degree that doing so is recognized to be sufficiently beneficial.[7] Though the good of society is important, it potentially can justify doing anything to certain individuals, no matter how destructive, unless some standard of human dignity prevents that from happening. The reason for this outcome is that according to a utilitarian perspective, what ultimately matters is the benefit itself (e.g., pleasure or preference satisfaction), not the individuals who benefit. Individuals whose existence imposes a burden on the whole—whether that whole is understood, for instance, in terms of society or of the family—must be eliminated to improve the well-being of the whole. At the least, individuals are expendable if profits or knowledge sufficiently beneficial to many can be obtained at their expense. Not only Nazi Germany but also U.S. slavery and some of the most ruthless and destructive experiments on human beings in the United States and elsewhere have been justified by this ethical outlook.

Others not particularly well disposed to the notion of human dignity reject its high regard for freedom of choice or bodily integrity—although they have been persuasively critiqued for doing so. Those most skeptical about freedom of choice include some in the social and biological sciences. Some psychiatrists and psychologists in the Freudian school, for example—even if not Sigmund Freud himself—argue that freedom of choice is an illusion. Choices are driven largely by unconscious and irrational forces.[8] Behaviorists who follow B. F. Skinner see such freedom as illusory as well, because behavior in their view is driven more by environmental stimuli than by freely willed choices.[9] Some biologists influenced by sociobiologists such as E. O. Wilson are similarly skeptical about attributing any special dignity to humans, because they are less impressed by any apparent differences between the abilities of people and of animals to make free choices than they are by biological similarities between humans and animals.[10] Such similarities go beyond the ability to experience pleasure and pain to encompass certain genetic, physiological, and other mental similarities.

Those skeptical instead about the high regard for bodily integrity in the notion of human dignity include so-called postmodernists and posthumanists. Postmodernists reject the "modernist" notion of a universally binding objective truth that has a wide range of implications for how people should be treated. Many postmodernists in the line of Richard Rorty would characterize as oppressive the idea that certain applications of technology to the human body such as cloning or radical genetic enhancement are inherently unethical (i.e., violations of human dignity).[11] One prominent option for such postmodernists is instead to espouse a radical autonomy whose focus on enabling the self to pursue what it wants and jettisoning other ethical standards is potentially quite destructive, as explained in later chapters of this volume. Posthumanists such as Ray Kurzweil, in contrast, doubt the value of the human body itself.[12] Bodily form is seen as an accident of history, which eventually will be replaced through developments in cybernetics and artificial intelligence. Since the human being has no lasting significance in this outlook, human dignity is an illusion.

Despite the fact that such outlooks as those just mentioned do not have a satisfactory basis for affirming human dignity, it is interesting to note that their adherents typically are reluctant to jettison the notion of human dignity entirely—perhaps because of its intuitive importance. For instance, Rorty opts for a form of moral obligation rooted in "a sense of solidarity with all other human beings."[13] Accordingly, some observers claim that recognition of human dignity in some form is unavoidably a part of every approach to bioethics.[14] Though that may be an overstatement, and the case for human dignity must still be made, the claim underscores the centrality of human dignity in bioethics.

Dignity Rooted in Human Characteristics

In the face of various challenges, then, a widely shared commitment to human dignity has persisted: the conviction that human beings, for some reason, have a special worth that warrants respecting and protecting them. Human dignity is not, as Ruth Macklin and others have claimed, the same thing as respect for persons. Rather, it is the basis for why such respect is warranted.[15] The big question is, What is that basis? Many people have addressed this question, and their responses are basically of two types. One roots human dignity in specific human characteristics. The other roots it in the simple fact of being human.[16]

The first type of response maintains that human beings have dignity because of one or more characteristics that are typically human. This view can be traced back as far as Marcus Aurelius and even earlier Stoic philosophers, who held that human beings have a basic equality rooted in their common ability to reason. It can be spotted occasionally in later periods—for example, in Renaissance thinkers such as Pico della Mirandola and Enlightenment philosophers such as John Locke. A classic statement of a form of this outlook is found in Blaise Pascal's summary observation that "all our dignity consists then in thought."[17]

A full-blown account of human dignity rooted in reason, however, found its most complete form in the work of Immanuel Kant, especially in his *Groundwork of the Metaphysics of Morals.* There he argues that "morality, and humanity so far as it is capable of morality, is the only thing which has dignity."[18] In other words, human beings do not have dignity simply because they are human but because and to the extent that they are capable of morality. Because for Kant "morality lies in the relation of actions to the autonomy of the will," he concludes that "autonomy is therefore the ground of the dignity of human nature."[19] Simply put, human beings have dignity because autonomous reason rather than impulses or the pursuit of personal or social benefit governs their actions.[20]

According to Kant's principle of autonomy, a human being "is subject only to laws which are made by himself and yet are universal."[21] Both parts of this principle are essential for Kant. Moral decisions must be self-made rather than imposed by others, even by God; but they also must be decisions that could be made and acted on by everyone consistently rather than products of an individual's personal view of reality (as in postmodern autonomy). In Kant's words, "all merely relative" ends are excluded: "The principle of autonomy is 'Never to choose except in such a way that in the same volition the maxims of your choice are also present as universal law.' "[22] Because they have autonomy, human beings have dignity, as opposed to price:

> Everything has either a price or a dignity. If it has a price, something else can be put in its place as an equivalent; if it is exalted above all price and so admits of no equivalent, then it has a dignity.[23]

Accordingly, human dignity requires that a human being be treated "never merely as a means" but "always also as an end."[24]

The legal philosophers Deryck Beyleveld and Roger Brownsword, among others, have tried to go beyond Kant and develop a reason-based approach to human dignity together with its implications for bioethics.[25] They affirm Kant's attempt to root human dignity in people's reason and capacity to be moral agents, but they prefer to follow the moral philosopher Alan Gewirth in adopting an understanding of agency that is focused more on choice.[26] For Beyleveld and Brownsword, "the essence of the dignity of agents resides in their capacity to choose, to set their own ends."[27] Consequently, they prefer to see human dignity more as empowerment than as constraint. Whereas

Kant's emphasis on people as "ends in themselves" fosters significant attention to limits on how people may be treated, even when they are acting on themselves, these authors see protecting each individual's right to choose as the primary mandate flowing from rooting human dignity in reason.

Despite the preoccupation with individual rights in many discussions of human dignity, especially in the West, the focus on the individual as opposed to the community is not inherent in the concept. A communitarian approach can champion human dignity in various ways. For example, it can establish respect for autonomy and choice as the hallmarks of what should characterize a society. Conversely, it can promote a vision of how people should and should not be treated that limits individual choices.[28]

Critical Assessment Criteria: Comprehensiveness, Consistency, and Credibility

Regardless of its individualistic or communitarian bent, what are we to make of any attempt to root human dignity in human characteristics such as reason and autonomy? To answer this question, it is helpful to consider three criteria: comprehensiveness, consistency, and credibility. *Comprehensiveness* is important because we want to make sure that our understanding of human dignity covers all the people to whom the term appropriately applies. Proponents of competing outlooks recognize that there is a problem if one's understanding of human dignity does not apply to all who are human. *Consistency* is important because our approach to human dignity must be able to withstand the critiques it levels at other approaches. *Credibility* is important because, when all is said and done, we want an understanding of human dignity that is plausible—one that accords with what we know about the present and what we hope about the future.

The first criterion, then, is comprehensiveness. Are all human beings accorded human dignity in this first conception of human dignity? Simply put, the answer is no, because specific human characteristics can be compromised or absent in particular human beings. If having human dignity requires possessing the ability to currently exercise moral capacity or autonomy, for example, then those who have mental disabilities, are comatose, are children, or are still in the womb do not have human dignity even if they are recognized as human beings. Often these are the individuals who are most in need of the protection that a concept of human dignity is designed to give.

Proponents of autonomy-based approaches have tried to give at least partial status and protection to those human beings in various ways. For example, Gewirth ties the level of a being's moral status to the degree to which that being has the necessary characteristic(s).[29] However, if human dignity is something one either has or does not have—as is typically affirmed—and if autonomy is the characteristic on which human dignity is based, then anyone without true autonomy does not have human dignity. Beyleveld and Brownsword agree, but they consider it possible to grant those in view moral status on the basis—and to the degree—that they *might* be moral agents complete with autonomy.[30] However, where there is a significant possibility that beings with autonomy are present, many people would consider it better to recognize and fully respect their human dignity. The strategy of Beyleveld and Brownsword seems unacceptably farfetched—they would consider even the simplest of life forms possibly to be fully autonomous beings and therefore would give them some sort of partial respect.

The second evaluation criterion is consistency. Can this approach to human dignity escape its own critique? Simply put, the answer again is no. The focus on characteristics itself is vulnerable to the very criticism it makes of its alternatives: It reduces human beings to what people in general or a particular community value about them, so in principle it invalidates ascribing human dignity to them. To put this in Kantian terms, like "things" that have "price" rather than "dignity," human beings also have a price or value that they as rational moral beings bestow on themselves because of some characteristic.[31] That a particular characteristic such as moral capacity or autonomy is uniquely valuable and therefore sufficient for granting human beings the exalted status of "human dignity" may seem intuitively plausible to many, but it does not seem so to others. So honesty requires us to classify this characteristic—and thus the human being who has it—in terms of price or value rather than dignity.

This brings us to the third evaluation criterion, credibility. How credible is the idea that the basic significance of human beings is reducible to one or more particular characteristics? Kant, for example, has been criticized for reducing what ultimately matters about human beings to the mind—to the *rational.* That outlook demeans bodily existence, which is so important in matters of bioethics. As the biologist, physician, and social thinker Leon R. Kass has argued, along with many others, "The account of human dignity we seek goes beyond the said dignity [of rationality] to reflect and embrace the worthiness of embodied human life, and therewith of our natural desires and passions, our natural origins and attachments, our sentiments and aversions, our loves and longings."[32] Postmodern outlooks narrow the focus even further in their devotion to radical autonomy and choice. Their preoccupation with process leaves out of account something very important. "Everything will depend, finally, not just on the possibility of choice, but on what is chosen," notes Kass, warning us to distinguish "cleverness about means from wisdom about ends."[33]

Limitations and Assumptions

Human rationality and autonomy are not the flawless dignity-grounding characteristics some in effect assume them to be. Human beings can indeed typically reason and make choices, but how flawless are their mental capacities? It would be great if these capacities were always *accurate.* But are they not in fact ever subject to limited and sometimes mistaken information? It would be wonderful if our mental capacities were always *objective.* But are they not in fact *self-centered*—employed so often for personal gain? It would be tremendous if these capacities were always *healthy.* But are they not in fact subject to the limitations of illness?

The credibility of an approach to human dignity that depends on idealized but unrealistic characteristics of human existence is therefore limited. It is no wonder that some who see dignity as tied to human mental capacities conclude that such dignity is a fiction: The proposition that "all men-qua-men are possessed of dignity . . . is indeed 'a fact' denied and invalidated by the greater part of our experience and knowledge," avers one Kantian scholar.[34]

In sum, then, approaches to rooting human dignity in specific human characteristics do not fare particularly well in meeting the criteria of comprehensiveness, consistency, and credibility. Much of the reason for this is that such approaches, as one secular critique puts it, are "based upon an anthropological 'creed'—not necessarily a religious creed."[35] The assumptions they make about human beings are as much leaps of faith as their religious counterparts typically risk. Creeds may be faulty, whether religious or not. So religious approaches should not be disadvantaged per se in discussions of human dignity. Rather, their comprehensiveness, consistency, and credibility should be assessed in the same manner as other approaches. All approaches are based on creeds that must hold up under scrutiny. There is an alternative creed that grounds human dignity in a very different way, and we will turn to that now.

Dignity Rooted in Being Human

If we recognize the importance of human dignity but are persuaded that basing it on the possession of particular human characteristics is not sound, we might consider the possibilities of rooting human dignity in being human per se. How could we anchor this common nature and meet our three criteria? One way is to focus on a source from which *all* characteristics may be said to flow, such as the human genetic code. UNESCO's 1997 *Universal Declaration on the Human Genome and Human Rights*, for example, affirms that all human beings are equal in dignity because of the underlying unity provided by the human genome.[36] Though this commonality may suggest the basic equality of all human beings, it leaves unaddressed the significance of all human beings.

Another way to avoid relying on particular characteristics is to consider the entire complex of human capabilities—such as reason, moral choice, language, and emotions—as an inseparable whole that provides the basis for human dignity. The international political economist Francis Fukuyama, for instance, takes this approach but admits that exactly what makes the whole sufficiently distinctive to warrant according it human dignity is "mysterious."[37] As in the case of the genetic code, this approach provides a way to distinguish the human from the nonhuman. However, "mystery" (or, for that matter, our self-beneficial preference for more complex human characteristics over simpler nonhuman ones) falls short of providing an adequate basis for the elevated significance typically attached to bearers of human dignity. Moreover, even understood as a package, a set of human characteristics can vary from person to person so substantially that comprehensiveness and credibility problems similar to those discussed above cannot be avoided.

If people's significance cannot be rooted in these specific sets of characteristics, then we must consider whether it can be found in something or someone beyond themselves, even if some degree of "mystery" is unavoidable at this point. One candidate would be the sort of universal force acknowledged in Buddhism.[38] Because that force is in all living things, however, whatever dignity it imparts is not particularly human. Yet if there is a God who establishes a special relationship with human beings that confers special worth on them, all people may be said to have a dignity that is distinctively human. No such account of human dignity has had greater influence than the one portrayed in the authoritative writings of several major religious traditions, where human beings are described as created in the "image of God." In addition to its role within religious traditions such as Judaism, Christianity, and Islam,[39] this account has had a substantial impact on public formulations of the concept of human dignity.[40]

The Image of God as Status and as Standard

A particularly well-developed and influential version of this understanding of human dignity tied to human creation in God's image is found in the Christian Bible (the quotations that follow are from the New Revised Standard Version). Two basic terms express the idea of "image": the Hebrew *tselem*/Greek *eikon* (generally translated as "image"), and the Hebrew *demut*/Greek *homoiosis* (generally translated as "likeness"). Although there have been historical attempts to distinguish these two terms, it is generally recognized that the two are used almost synonymously throughout the Bible.[41] Usually one or the other appears, but occasionally, as in the account of the original creation of humanity in Genesis 1, both are employed. The sense conveyed is that of an image truly representative of God.[42] Human dignity in this view is not tied to a claim that human beings are divine or inherently worthy apart from God, nor is it a function of human autonomy independent of God whereby people assume the authority to declare their own worth. Instead, human dignity is grounded in humanity's unique connection with God, by God's own creative initiative.

Much discussion of the image of God is misleading, because it either fails to distinguish the two different meanings of the concept or applies the concept inappropriately. The image of God can refer to either a status or a standard. As a status, it refers to the dignity that all human beings have because of their creation by God. All human beings have this in the beginning (at the Creation—Gen. 1:26, 27; 5:1), throughout history (after the Fall—Gen. 9:6; 1 Cor. 11:7; James 3:9), and at the end (when Christ returns—1 John 3:2). Although most of these passages affirm that human beings *are in* the image of God, the status in view is so concrete that the biblical text can also affirm that human beings *are* the images of God (1 Cor. 11:7). Nevertheless, each of these passages is talking about something that is fully true of every human being. This status is not related to any characteristics or capacities that people have or lack to some degree.

Conversely, other biblical passages, found only in the New Testament, refer to the image of God in terms not of what human beings already are—their status—but of a standard for how human beings are to be and live. This standard is rooted in God's own character and will. The standard is what God intends the believer to be (Eph. 4:24) and be conformed to (Rom. 8:29)—it is the goal of the renewal (Col. 3:10) and transformation (2 Cor. 3:18) that a believer is experiencing. Whereas all human beings have the status of images of God, human beings vary considerably in the degree to which they measure up to the standard of the image of God.

A common feature of both the image as status and the image as standard is that they are thoroughly positive and uncompromised. Accordingly, theological claims that the image of God was "tarnished" or "diminished" after the Fall are biblically inaccurate and therefore misleading. Nowhere does the Bible indicate that either the image as status or the image as standard can be compromised. Theological suggestions to the contrary can have dangerous (though perhaps unintended) consequences. Such is particularly the case when the image as status is not distinguished from the image as standard, and the status of people is thought to depend on their achieving certain standards—that is, having certain characteristics. Those least measuring up are at risk of being accorded less dignity and thus given less protection and empowerment, potentially to their great peril.

A biblical understanding of human dignity rooted in people being the images of God avoids this outcome both by resisting the notion that the image of God was altered by the Fall and by refusing to attach the status of the image to particular human characteristics or capacities. Accordingly, it is not surprising that the concept of the image of God does not even appear in the biblical text after the opening (nine) chapters of Genesis until the New Testament. The Bible is concerned to establish the status of the human being *created* in God's image, and it is further concerned to establish the standard of the human being *re-created* in God's image in Christ. Both warrant further consideration here. The first has special significance for human dignity as an ethical standard, the second for human dignity as an ethical virtue.

Creation

In terms of creation, Genesis 1 (with a reaffirmation in Genesis 9) indicates that the image of God attaches to that which is human as opposed to animal or plant. As a human child was considered the *tselem* of a parent (Gen. 5), and a *tselem* in the ancient Near East could refer to a statue reminding people of a king's presence, human beings were created to have a special, personal relationship with God that includes their being God's representative in the world.[43] Accordingly, the Bible speaks of people not only as being *in* the image of God but also as *being* the image of God. This is striking, because images of God are strictly forbidden in the Bible (e.g., Deuteronomy 4). Yet the consistent message is that people are not to fashion images to make God the way they want God to be any more than they are to try to be God themselves. They are to manifest God to the world in accordance with the way God has made them and continues to direct them to be.

In sum, these passages about human beings being created in God's image are talking about their status. But while the standard of God's image and what it requires of human beings can always be assumed throughout the Bible, it is not addressed in these passages—nor are the normative human characteristics or capacities that it entails. And so scholars' attempts to define humanity-as-God's-image in terms of specific God-like characteristics, such as the ability to *rule*, *reason*, or *relate*, are misguided.[44] They are reading into the biblical text rather than reading from it. For instance, the idea that the human being is intended to rule creation (i.e., rule it rightly, exercising faithful stewardship over it[45]) is biblically sound, as the Psalmist observes to God: "You made him ruler over the works of your hands" (Ps. 8:6). It is even accurate to recognize that this ruling is closely associated with being created in the image of God. In the very next verse, after the text records the creation of the human being in the image of God, it reports God's instruction to the first humans to rule over every living creature (Gen. 1:28). But this latter instruction is not part of the description of what creation in God's image is—it is a separate matter that exemplifies what should be expected of one who is created in God's image.

We should similarly expect to find rational abilities in those created in God's image. In fact, the particular rational and specifically spiritual capacities found in human beings are among the most evident distinguishing features of human beings as opposed to animals. However, to define the dignity of human beings created in God's image in terms of these capacities is to take on all the problems resulting from the Kantian approach of making human dignity ultimately dependent on particular human characteristics. Once characteristics rather than human beings themselves are the locus of value, the drift begins toward the commercialized dehumanization of the final secularized society, Babylon, where everything is up for sale according to the value attached to its characteristics (whether material or nonmaterial): "cargoes of cinnamon and spice, of incense, myrrh and frankincense, of wine and olive oil, of flour and wheat; cattle and sheep; horses and carriages; and bodies and souls of people" (Rev. 18:13).

With the biblical text and all the scientific and other descriptive tools available to us today, we can say a great deal about who human beings are. In anticipation of the return of Christ, the perfect image of God (2 Cor. 4:4; Col. 1:15; Heb. 1:3; cf. John 14:9), we can speculate about the ways we will find we are in the likeness of Him when we stand before Him. But the Bible itself warns us that while "we know that when he appears, we shall be like him," the specifics of what that likeness will be are not now known (1 John 3:2). The implication is that we do better to affirm what we can know—that is, the fact that human beings are the images of God—than to speculate about what we cannot know. Accordingly, although various characteristics are attributed to human beings in the biblical text, they are never identified there as what constitutes that image. To complicate matters further, angels appear to have most of the capabilities (memory, will, moral and spiritual capacity, etc.) that some would identify as constituting the image of God, but they are never identified as being created in God's image. The picture presented in the biblical writings is that human beings themselves—not some particular attributes or functions—are through God's creation the image of God.

The capabilities of angels also cause difficulties for the third common approach to the image of God: defining that image in terms of people's relational capacities. Angels, too, appear to have the capacities to relate to God and others. Again, the association of the image of God and such capacities is understandable. The very same biblical verse (Gen. 1:27) that records the creation of the human being in the image of God then reports the creation of human beings as male and female. However, these are two distinct affirmations, and there is no indication in the text that one of these affirmations defines the other one. In fact, what the close proximity of the two affirmations calls to the reader's attention is their significant difference. Creation in God's image applies to the human being per se (the singular pronoun—often translated "him"—is used), whereas creation in two genders applies to human beings in their plurality of differences (the plural pronoun "them" is used): "In the image of God, God created him; male and female, God created them." Human beings are different from one another—which makes relationship more fulfilling—and analogies to the trinitarian nature of God are appropriate to ponder. Nevertheless, the biblical text does not connect the image of God to human beings in the context of their differences—or the relational capacities connected with those differences—but to the human being per se.

The significance of the Fall for the image of God does not rest in any damage done to the image, as explained above, but rather in the fact that those created with the status of God's image no longer measure up to the standard of what God's image should look like—even though they retain the status of being the images of God. As the Bible describes this predicament, people decided to do things their own way, to give in to the temptation to "be like God" on their terms rather than God's (Gen. 3). As a result, they have become alienated not only from God but also from their own best selves, other people, and the rest of creation. Their abilities to rule, reason, and relate well have been damaged severely (Ps. 14, expanded in Rom. 1, 8), and people now seek to create images to worship (including themselves) because they have lost sight of the fact that they are images of God created to reflect and direct worship toward God rather than to be worshiped themselves.

Even in this alienation, human beings retain their full status as images of God. In this we see the first evidence of God's redemptive intentions concerning humanity—the continuing of God's identification with humanity.[46] The ethical standard of respect for human dignity gains its force precisely from this ongoing connection with God, for those who are dealing with human beings are dealing in a significant sense with God. Killing an innocent human being is destroying an image of God without warrant from God and for that reason is unacceptable (Gen. 9:6).[47] Equally unthinkable is the attempt to demolish a human image of God verbally through cursing (James 3:9). It is not surprising that Jesus connects murder with attempts to destroy people verbally and condemns them both (Matt. 5:21–22), because both are designed to tear down an image of God.

Re-Creation

The ethical standard of respect for human dignity rooted in the biblical account of Creation, as noted above, is affirmed in various religious traditions. So is the virtue of recognizing the dignity of human beings in words and actions, along with the difficulty of doing that once one is alienated from God. What the remainder of the biblical story adds is a particularly Christian account of the re-creation that is possible in Christ. With that re-creation comes the ability to better live out the virtue of human dignity because one who has the status of being God's image lives more and more according to the standard of what the image of God should look like. For alienation to be replaced by reconciliation, according to this account, people literally must undergo a new creation (2 Cor. 5:17–19; cf. Eph. 4:24). They must recognize the hopelessness of their alienation, give up all attempts to improve their situation through their own (futile) efforts, and invite God to re-create them in the image of God-revealed-in-Jesus-Christ. Though the creation is new, the image on which it is based is not, for Christ is identified not only as the image of God but also as God who created humanity in God's own image in the first place (Col. 1:15–16).

This new creation is portrayed in both ontological and logical terms. It is an ontological fact in that it is an event in time with implications for one's being. One who is re-created in Christ is again created with the status of being the image of God. But the stranglehold of the alienating power of sin is also broken. That makes possible the logical effect of the new creation—that is, the process that flows logically from this event. People begin to live out in practice the standard of the image of God—a standard that their status as images of God called for all along. In that sense, they become in practice who they already are in being.[48]

The focus here is primarily on being the image oneself rather than on recognizing the image in another—even though both are legitimate concerns. It is all too easy to use the question of whether someone else is truly an image of God (and therefore worthy of our protection) to avoid addressing the question of what it means to be the image of God ourselves. Jesus redirected the person who asked "Who is my neighbor?" (Luke 10:29) to consider who in the Good Samaritan story "was a neighbor" (v. 36)—that is, acted as one. Those adept at being the neighbor—and the image of God—are more likely to excel at seeing the neighbor, recognizing the image of God in others and treating them appropriately.[49]

When people are renewed "in the image of their Creator," the result is described in terms of not only renewed individuals but also a renewed community: "Here there is no Greek or Jew, circumcised or uncircumcised, barbarian, Scythian, slave or free" (Col. 3:11). Differences no longer divide—they disappear or in some cases can even enhance community, in which the human dignity of all is recognized. Those who are renewed images of God warrant no better treatment than do any other human beings in this view, because all human beings have the full status of human dignity by virtue of their original creation in God's image. However, those who are renewed images are characterized as increasingly more capable of exercising the virtue of human dignity than they would be otherwise.

More could be said about human dignity in the biblical writings.[50] God makes covenants with human beings, became a human being in Jesus Christ, retains that humanity eternally, died in humanity's place to pay the penalty for human rebellion against God, and will appear personally to bring humanity into an unending celebration of life with God, when people finally will understand all that being in the likeness of God entails. All these historical developments fill out the biblical account of human dignity but also rest on the basis that human beings are images of God—arguably the very heart of what it means to be human.

Meeting the Three Critical Assessment Criteria

How does rooting human dignity in being human—as opposed to possessing specific human characteristics—fare according to our three major evaluation criteria: comprehensiveness, consistency, and credibility?

Under our first—comprehensiveness—by definition this approach recognizes the human dignity of all human beings. Human dignity does not depend on having certain characteristics that a particular human being may or may not have. From Genesis onward, the fact that a being is human rather than not human is what identifies that being as in the image of God. Today we can add the scientific knowledge that this distinction is present at the earliest embryonic stage, when the genetics of the embryo already distinguishes this being as human as opposed to something other than human. To be sure, being human is more than a matter of genetics, but it is not less. It includes the material and nonmaterial aspects that constitute human beings, but the fact that those aspects are present and have the capacity to develop further is signaled by the distinctive genetics and life present in even the least capable members of the species *Homo sapiens.* Genesis and genetics together include, under the umbrella of human dignity, all human beings—embryos and elderly, diseased and disabled.

All four of those groups have come under attack as potentially expendable not only from a utilitarian or postmodern or posthuman perspective but also from the modernist, rationalistic perspective represented by Kant. The Bible explicitly upholds the dignity of the diseased, the disabled, and the elderly, but does it do the same for the embryo? This question warrants brief comment here because of the number of emerging biotechnology issues that threaten the dignity of human beings at the earliest stage. Though the list of biblical passages with implications for the humanity of the human embryo is long, two passages stand out as particularly instructive, one in the Old Testament and one in the New Testament. Both are consistent with the Genesis-based view that a being who is identifiably human—whether by genetic or any other criteria—is an image of God whose human dignity must be honored.

The first such passage is Psalm 139, which opens with six verses on God's sovereignty—more specifically God's omniscience and omnipresence. The next six verses describe our natural bent to try to escape from God to find some situation in which we can do with, and to, our life whatever we want—if we go high enough, or low enough, or far enough. Verses 13 and following then bring home the fact that there is no such place—even in the womb—even at the stage when there is an "unformed body" (which some translators translate as *embryo*, because that is the stage before formation takes place). Even in the womb we are not far enough away from God—we are not instead in the realm of "private choice" or "science"—we are not welcome to do with a human being whatever we think will benefit us.

When we come to the first chapter of Luke, in the New Testament, this close connection between God and human-beings-as-God's-images becomes profoundly closer when the very God of the universe becomes inextricably and eternally connected to a human body in the person of Jesus Christ. Before noting the importance of this for embryonic human life, it is essential to note the more general significance of this for what constitutes a human being and, in turn, human dignity.

It is not uncommon to think of God exclusively in nonbodily terms and so unconsciously to consider what is important about human beings made in the image of God to be their nonbodily aspects—spirit, mind, reason, and so forth. It is easy, by implication, to downplay the significance of the body in determining what constitutes a human being. But this common view of God is not the view of the Bible. Admittedly, at first God existed apart from humanity and human bodily form. But when "the Word became flesh and dwelt among us" (John 1:14), an eternal identification with humanity was established, such that Christ still had a human body after his resurrection. Consider, for example, Luke 24:39, where Christ says: "Look at my hands and my feet. It is I myself! Touch me and see, a ghost does not have flesh and bones, as you see I have." Accordingly, later in the New Testament, it is affirmed of Christ's ongoing existence: "In Christ all the fullness of the Deity lives in bodily form" (Col. 2:9). And that is for eternity.

As suggested above, however, God's becoming a human being is significant not simply because there was an adult Jesus and a child Jesus but also because there was an embryo Jesus. In Luke 1, we see Mary becoming pregnant with Jesus and right after that conception—the text says "immediately"—she makes what was probably a several-days' journey to visit her relative Elizabeth, whose own baby leaps in Elizabeth's womb at the encounter both with Jesus as a several-day-old embryo and with his mother. In fact, Elizabeth addresses Mary as already Jesus's mother, emphasizing that not only is Jesus's presence already a reality but so are her role and responsibility as his mother.

The Incarnation began, not with Jesus as an adult or a child, but with Jesus as an embryo. It is hard to imagine God taking on some form that could have been discarded as a mere blob of material. To the contrary, the Bible sees the embryo, before formation in the womb has begun, as God's creation, in God's image, fully suited for God's own presence, fully possessing human dignity. A biblical view of human dignity, then, is truly comprehensive, including all human beings from the embryonic stage onward.

Under our second assessment criterion, consistency, the use of a biblical outlook on human dignity, like rooting human dignity, sometimes faces the same critique that it levels at other approaches. It is accused of promoting a form of "speciesism," a parallel to racism, that blindly elevates humans at the expense of others, particularly animals, thereby promoting the kind of exploitation of God's creation that those espousing a biblical approach criticize in other outlooks.[51] However, such inconsistency is not present in the biblical outlook itself—though some people claiming to represent it have undoubtedly been guilty of this inconsistency. The notion that human beings alone have a special dignity, by virtue simply of their being human, is indeed the biblical view. Such speciesism, if such it may be called, however, hardly has the negative implications of racism, either in general or with regard to animals in particular. Attempting to connect the speciesism of biblically based human dignity with racism overlooks a more obvious connection. It has been precisely this speciesism of all human beings having human dignity simply because of their humanity, not their characteristics, that has been the driving force behind the movement to *end* racism.

Moreover, to think that speciesism in a biblical context would sanction the unethical treatment of animals is to miss the heart of a biblical outlook on human dignity. Because human dignity so understood is rooted in creation in the image of God, it does not merely give human beings exalted status and the autonomy to do with the rest of creation whatever they wish. Such autonomy is what people have sought from the very beginning, when Genesis 3 says they were tempted to be "like God." There the explicit meaning of being like God was that they would gain knowledge—knowledge of good and evil. The implicit meaning is that they would gain autonomy—they would establish themselves rather than God as the ones to choose what was in their best interests. And so the temptation remains today, for all people. The temptation is to consider our rational capacities—our knowledge and our autonomous choosing—to be what gives us significance as human beings.

However, when the biblical Psalmist observes before God that "You made humanity ruler over the works of your hands" (Ps. 8:6), the logical implication is not that human beings are great; it is rather: "O Lord, our Lord, how majestic is your name in all the earth!" (v. 9). Such is the case because humanity is nothing in itself and would be nothing apart from God's doing: "What is the human race, that you are mindful of it . . . ?" (v. 4). A biblically informed view is not that human dignity gives people liberty to do what they like with the rest of creation but rather that as images of God they are bound to care for creation as God would care for it. Only as representatives of God in the world do they have significance or authority. God provides animals with the nourishment that they need and plants with a splendor all their own (Matt. 6:26, 29), for from the very beginning God has done what is necessary to ensure that the entire creation is "good" (Gen. 1:10, 12, 18, 21, 25). It is humanity living its own way, apart from God, that has ultimately caused the creation to "groan" as it does now (Rom. 8:22)—not people living as the images of God they were created to be.

This brings us to our third assessment criterion: credibility. We have already seen how a biblical outlook on human dignity effectively avoids the credibility problem of the approach that roots human dignity in particular human characteristics. It does not mistakenly reduce what is important about human beings to their rational or other specific capacities. Rather, while affirming the importance of all aspects of our humanity, including our bodily existence, it roots human dignity in the totality of our humanity in a way that is not jeopardized when one or more aspects of that humanity are compromised.

In a broader sense, however, a biblical approach to human dignity is more credible than the alternatives because it best accords with what we know about humanity in the present and hope for humanity in the future. We have already critiqued the unrealistically high view of rational human beings assumed in Kantian and utililitarian outlooks, and we have noted the low view of humanity driving the posthumanist alternative. If the rationally or otherwise weak can be weeded out of society in the first two outlooks, everyone human is ultimately in jeopardy in posthumanism. The postmodern outlook is capable of recognizing both the good and the bad in humanity, but the primacy of radical autonomy undermines the notion that there are any consistent ethical standards by which to distinguish good from bad or to foster one over the other.

By comparison, a biblical approach to human dignity has particular credibility. It takes into account the dynamic tension between human beings as profoundly limited yet profoundly significant. On the one hand, it is anything but naive concerning human nature. Not only are our knowledge and ability to understand significantly limited by our finiteness, but we are led astray by our self-centeredness. As summarized in Romans 3:10–11: "There is no one righteous, not even one; there is no one who understands." If our dignity depended simply on some capability that we have, we would be in trouble. But human dignity in biblical perspective does not depend solely on who we are but, more important, on who God is—as well as on what God has done, is now doing, and will do in the future.

Comparative Advantages

We cannot make credible sense out of our deep conviction that we have dignity and our lives are meaningful, simply on the basis of an empirical assessment of our flawed capacities and the outlooks that drive the other approaches to human dignity already mentioned here. But the biblical affirmation of our wonderful creation in God's image, our alienation from God as we pursue our own way, and the possibility of our renewal in Christ account well for our realistic disappointments over our limitations yet our conviction and hope that we can expect more.

Ultimately, those completely renewed in God's image will experience the full potential of all the wonderful capacities that were associated with humanity's original dignity as images of God. However, we live in the between-times when our dignity persists even though the excellence of all our capacities does not. If our dignity depends on such capacities—particularly on their renewal—it is no wonder that the posthumanists are trying to achieve perfection including eternal life in this world. However, a biblical perspective is not only more realistic but also very reassuring on this point. It presents God in Genesis 3 barring people from the eternal life that they would have had in this world had they chosen to be faithful images of God—had they chosen to be like God in the right way. They are barred because they instead went their own way and yielded to the temptation not so much simply to *be like* God as to *be* God. God's closing the door to eternal life in this present world was thus not merely a punishment but an act of great mercy. It actually opened the door to the possibility of the complete renewal that will occur for those whom God finds faithful at the end of their lives in this world—when human dignity and all human capacities will be rejoined completely.

Tasks for Bioethics: Weighing the Violations of Human Dignity

As has been implied here, people most commonly invoke human dignity in situations where the worth of human beings is brought into question when they are inappropriately used (i.e., demeaned), forced, or injured. The idea is that human beings should not merely be *used* because their dignity requires that they be treated as having intrinsic, not merely instrumental, worth. They should not usually be *forced* because their dignity mandates that their wishes be respected. They should not normally be *injured* (through someone's action or through their lack of access to basic life-sustaining resources) because their dignity entails that their well-being be preserved.

In some bioethical issues, one or more of these dignity-related concerns argue persuasively against other (often utilitarian) considerations, and both major approaches to human dignity agree that respect for human dignity must have priority. For example, in evaluating a form of human experimentation, people commonly insist on obtaining the informed consent of participants, lest participants' dignity be violated when something is done to them against their wishes. No amount of benefit to society warrants such a violation. In matters of resource allocation, human dignity may be invoked to argue that the allocation producing the greatest overall social benefit is not the right one if the burden that certain individuals must bear to bring it about is too heavy. Not only may some people be severely injured, but also the very process by which absolutely anything can be done to them if it results in greater benefit to society is demeaning. Human dignity also is rightly invoked to protest the injury involved in human cloning for reproductive purposes, as long as animal studies show that attempts to clone humans almost certainly would result in children with serious, even lethal, deformities.

In other bioethical debates, human dignity is not so unambiguously on a particular side of the issue. One reason for this is that people tend to invoke dignity intuitively and self-servingly rather than after distinguishing and prioritizing the three violations of human dignity noted above. There is good warrant, however, to consider injury to be, generally speaking, the most serious violation of human dignity, followed by using people, and followed in turn by forcing people. Whereas all three forms of human indignity are typically involved in injury, using people normally involves only two forms (using, and forcing that comes with using people in a way that they would not choose), while forcing often involves no indignity other than the limiting of someone's choices because morality requires it. To be sure, some limitations of freedom are morally disturbing, in part because they are violations of human dignity. Nevertheless, not only does injury often compound such a violation, but injury is a particularly significant violation of human dignity when it is fatal. In such a case, the killer uses (i.e., destroys) all that someone is in order to gain something much less significant, and the loss of the deceased person's freedom is total.

For these reasons, injuring, especially killing, is sometimes considered to be an ethical violation separate from—and usually more serious than—violations of human dignity. However, such an understanding differs little in its implications from an understanding that sees killing/injury as the most objectionable form of indignity—for example, the Bible's view of killing as destroying an image of God (reflected in Genesis 9:6).

Some of the controversy over the implications of human dignity, then, is due to a lack of clarity over its different aspects and how to prioritize them. Some of the controversy, however, is also due to the fact that more than one anthropological creed is influential, leading to competing conceptions of human dignity. Sometimes the clash involves a conflict between those concerned about injuring people and those concerned about forcing people. As noted above, those who root human dignity in a particular characteristic such as reason not only tend mistakenly to weigh reasoning or choice as the highest value to respect in a situation, but they also tend to prioritize unfairly the interests of those best able to reason.

In the debate over embryonic stem cell research, for instance—a process in which human embryos are killed to obtain embryonic stem cells—people who consider the freedom to choose as definitive for human dignity often see no conflict. When a human embryo, a researcher, and a patient needing a new treatment developed are in view, there are only two human beings with the ability to choose present, so the researcher's and patient's wishes prevail. Opponents of embryonic stem cell research with a different, more biblical view of anthropology instead hold that three human beings are present. Accordingly, they see the situation as a conflict between two affronts to dignity, in which a greater violation would be done by fatally injuring the embryonic child than would be done by forcing researchers to limit their investigations to stem cells that can be obtained without killing human embryos.

In end-of-life situations, where there is no doubt about what human beings are present, the clash over whether to prioritize forcing or injuring is more straightforward. Some maintain that death with dignity requires having the freedom and ability to end one's life whenever one wishes, even if someone else has to do the killing. Others maintain that killing oneself (as in assisted suicide) or killing another (as in euthanasia) is the ultimate injury and must not be supported or allowed. Though one could simply recognize fatally injuring someone as the greater indignity involved here, there are additional dignity considerations that argue against assisted suicide and euthanasia. For example, to disconnect freedom of choice from—and elevate it over—the ethical requirement of respect for human life is to jeopardize that very freedom. As experience in The Netherlands has demonstrated, once killing patients who request death is accepted, it is predictable that decisions will next be made to kill some incompetent patients who cannot express their wishes but are in similar medical conditions. And finally decisions will be made to kill some competent patients without obtaining their consent, again because they are in similar medical conditions.[52]

Other bioethical debates are even more complicated in that two elements of human dignity—preventing people from being injured and preventing people from being used—are in conflict with a third element: preventing people from being forced. For this reason, the groups of people on each side of these debates are not the same groups as those in the debates mentioned above.

For example, in the debate over germline genetic intervention to enhance future generations of human beings, those who see the central issue as the limitation of people's choices tend to favor giving parents and society freedom to pursue such avenues. Postmodernists, for example, typically espouse this view, whether or not they couch their position in terms of human dignity. Others, including those who are biblically influenced, tend to be more concerned to protect people against significant injury, even if choices are limited in the process. They identify a threat to human dignity in subjecting young human beings to such procedures when the potential negative effects of genetic alterations for enhancement purposes are not well understood. That opposition is strengthened for many—not just for those biblically oriented but also for others including Kantians wary of utilitarian abuse—who see far more involved here than just the potential injury involved. They also recognize that the people doing the enhancement unacceptably use other human beings by altering them to exhibit traits that parents or society may find beneficial but that the ones who are altered may not.

Human cloning is a violation of human dignity in the same way. Cloning imposes on children a particular genetic code of someone else's choosing while denying these children one of the two genetic parents noncloned children are born with. The genetic imposition may be well intentioned in terms of particular desirable traits to be passed on. In the process, however, entire genomes of other genetic instructions are imposed, which will necessarily contain many other harmful elements. Who but the persons-to-be can say how unacceptable these other harmful elements are—not to mention how beneficial they consider the supposedly desirable traits to be? But these persons cannot consent because they do not yet exist. Forcing some people to limit their reproductive options, all else being equal, is not ethically attractive. But it is warranted by the greater violation of human dignity involved in both using and forcing other people by bringing them into the world through cloning. The indignity involved is only made worse by the likely injury, even death, of the cloned embryo due to the technical limitations of the cloning process.

Conclusion

Human dignity, then, plays a significant role in many bioethical debates. The language of human dignity is persuasive to many, and it can provide welcome common ground for mobilizing consensus in the public arena. At some point in certain debates, however, when concepts of human dignity irreconcilably conflict, it will be necessary to look at the relative merits of those concepts. In such discussions, no concepts should be excluded because they are religious, politically incorrect, or otherwise out of public favor. As we have seen, every concept is based on a creed—an anthropological creed—and all creeds must be held accountable to basic standards such as comprehensiveness, consistency, and credibility. According to such standards, a biblical approach to human dignity—rooting it simply in being human—fares well compared with other outlooks.

Of course, not all will agree. One scholar suggests the basic alternative to the biblical vision: "Our lives are the chance end products of blind evolutionary processes in an unconscious universe indifferent to our individual existence and inevitable demise."[53] Inspiring? A solid basis for the kind of human dignity that we know, deep down, to be real and precious? How much better are the words of the Psalmist (Ps. 100:3–5):

> Know that the Lord is God. It is God who has made us, and we are God's. . . . Give thanks to God and praise God's name. For the Lord is good; the Lord's love endures forever; the Lord's faithfulness continues through all generations.

CHAPTER FIVE

Biotechnology and the Quest for Control

> Men ought not to play God until they learn to be men, and after they have learned to be men, they will not play God.
>
> —Paul Ramsey, *Fabricated Man*

CURRENT and future technocratic "freedoms" advance elimination of less-than-perfect human bodies as a primary goal. This dynamic is worthy of careful examination. The purposes of this chapter are threefold. First, we measure some contemporary projects that deny human finitude and promote, instead, a technocratic agenda of full control of the human body—of nature itself. Second, we argue that these projects promote an ideal of human sameness by construing the stuff of human life as what we can manipulate according to our culture's standards of what is excellent, good, beautiful, and important. Third, and if anything even more frightening, we maintain that the project of genetic control and manipulation now under way requires elimination of bodies deemed unworthy, whether after conception but before birth through the abortion of "imperfect" fetuses or—and ideally under this agenda—by manipulating the genetic possibilities for a good "product" as part of the process of conception itself. These themes are woven together throughout the chapter.

The issues raised here are very complex. Thus many of the themes taken up cannot be addressed in the length and detail they deserve. We aim instead to heighten our collective attentiveness to certain pervasive themes and to demonstrate, however briefly, what seems to be at stake in how we think and act with respect to the technocratic impulses endemic to our culture.

Examining the Quest for Control over the Human Body

In our fast-paced, fitness- and youth-oriented culture, perfecting the human body has become a messianic project. The situation we face is this: Bodies are thought of increasingly as the exclusive property of an individual to do with as one sees fit. Bodies are also construed as malleable and "constructable." We are all enjoined—by advertising, cultural imagery on television and in films, science joined to profit in the biotechnology industry, popularizers of the genetic revolution, and others—to "get with the program," to hop on board and not remain stuck in superstition urging restraint or even curtailment of genetic and biological engineering.

It is easy enough to understand uncritical identification with the currents of one's own time because so many of them speak to real human needs, fears, and desires with their associated goods. All of us are bound up with some ideal of freedom and with the prevention of suffering. Therein lies a major part of the problem, at least if one follows Martin Luther's lead. Luther insisted that all our needs are bound to be distorted given human rebellion against God, beginning with the disobedience that got Adam and Eve thrown out of the Garden. Ever since, human life has been marked by a trace of this original willfulness, or so argued the great reformer. We are separated from God, the source of undistorted love. As well, given that Christian theological anthropology presumes intrinsic relationality—and that there is no primordially free self—sifting our cherished and essential commonalities (and "commune-alities") from unthinking absorption into dominant cultural forces is a delicate matter. This chapter provides examples of cultural acquiescence in ever-more radical manipulations of the human body.[1] An overarching and framing thematic of contemporary American culture, as noted, is a flight from finitude undermining recognition of the complexities and limits as well as the joys of embodiment—the givens of human life itself.

Luther's mordant view of the lingering implications of human defiance of the Creator requires fleshing out, beginning with reminders about the nature of Christian freedom and the fact that we are both creatures and creators. As creatures, we are dependent. It follows that our creaturely freedom consists in our recognition that we are not abstractly free, but free only in and through relationship. A limit lies at the very heart of our existence in freedom. Christian freedom turns on recognition of the limits to freedom. Where one draws the line between the good and ill uses of freedom is no easy matter, however. Theologians and philosophers have struggled with this for centuries. Surely our embodied natures help us to think about limits. Does a project promise a transhuman or posthuman future, as many currently do? Or is it one that recognizes our freedom to use our God-given capacities in ways that are always mindful of human limits and imperfections, of our status as both begotten natals and mortals?

In *Creation and Fall*, the German theologian and anti-Nazi martyr Dietrich Bonhoeffer frets that human being as creator easily transmogrifies into a destroyer as he (or she) misuses freedom.[2] There is a big difference between enacting on the one hand human projects as co-creators respectful of a limit because, unlike God, we are neither infinite nor omniscient, and by contrast on the other hand those projects that demand that humans embrace God-likeness for themselves to the point of displacing God Himself. With God removed as a brake on human self-sovereignty, we see no limit to what human power might accomplish. An alternative to this project of self-overcoming is an understanding of a humbler freedom, a freedom that never aspires to the absolute. This freedom is constitutive of our natures. The theologian Robin W. Lovin helps us to appreciate a specifically Christian freedom that is not opposed to the natural order but rather acts in complex faithfulness to it.[3]

One begins by taking human beings as they are, not as those fanciful entities sometimes conjured up by philosophers in what they themselves call "science fiction" examples, such as Judith Jarvis Thomson's famous abortion analogy of a woman forcibly hooked up to a violinist.[4] To be sure, the freedom of a real, not a fanciful, human being means among other things that, as Lovin puts it, one can "project oneself imaginatively into a situation in which the constraints of present experience no longer hold."[5] One can imagine states of perfection or near perfection. At the same time, actual freedom is always situated; it is not an abstract position located nowhere-in-particular. Freedom is concrete, not free floating. Freedom is a "basic human good," asserts Lovin. "Life without freedom is not something we would choose, no matter how comfortable the material circumstances might be."[6] Our reasoning capacity is part and parcel of our freedom. But that reasoning is not a separate faculty cut off from our embodied selves; instead, it is profoundly constituted by our embodied histories and memories.

Christian freedom, in Lovin's words, consists in our ability to "avoid excessive identification with the surrounding culture, since that tends both to lower . . . moral expectations and to deprive [persons] of the witness to alternative possibilities."[7] If the horizon lowers excessively, the possibility that we might exercise our capacity for freedom is correlatively negated. So the denial of freedom consists in part in a refusal to accept the freedom that is the human inheritance of finite, limited creatures "whose capacities for change are also limited, and who can only bring about new situations that are also themselves particular, local, and contingent."[8] To presume more than this is also problematic, launching us into dangerous pridefulness, often of course in the name of great ideals, such as choice or justice. So our freedom is at one and the same time both real and limited.

Projects of Self-Overcoming

Against this backdrop, let us examine several contemporary projects of self-overcoming that involve a negation of (or an attempt to negate) finitude and that rely on an uncritical endorsement of dominant cultural demands.[9] Such projects, remember, are tricky to approach critically because they present themselves to us in the dominant language of our culture—*choice*, *consent*, *control*—and because they promise an escape from the vagaries of the human condition into a realm of near mastery. Consider the fact that we are in the throes of a structure of biological obsession underwritten by pictures of absolute self-possession.[10] We are bombarded daily with the promise that nearly every human ailment or condition can be overcome if we just have sufficient will and skill and refuse to listen to any entreaties from critics, who are invariably portrayed as negative and against progress. For those whom the Canadian political philosopher of modernity Charles Taylor calls the "cultural boosters," our imperfect embodiment is a problem that must be overcome. For example: a premise—and promise—that drove the Human Genome Project (HGP), the massive mapping of the genetic code of the entire human race, was that it would enable us one day to intervene decisively to guarantee better if not perfect human products.[11] Claims made by the promoters and advocates of the HGP have, from the beginning, run to the ecstatic.

Take, for example, Walter Gilbert's 1986 pronouncement that the total human genome sequence "is the grail of human genetics . . . the ultimate answer to the commandment 'Know thyself.' "[12] In the genome-triumphalist camp, they are now talking about "designer genes." Note the following advertisement that had appeared in college newspapers all over the country and was reported by the *New York Times* on March 3, 1999:

> EGG DONOR NEEDED / LARGE FINANCIAL INCENTIVE / INTELLIGENT, ATHLETIC EGG DONOR NEEDED / FOR LOVING FAMILY / YOU MUST BE AT LEAST 5′10″ / HAVE A 1400+ SAT SCORE / POSSESS NO MAJOR FAMILY MEDICAL ISSUES / $50,000 / FREE MEDICAL SCREENING / ALL EXPENSES PAID.[13]

Only a few years ago, this was shocking. Now it is so pervasive as to be commonplace. As *Commonweal* noted in its editorial occasioned by this advertisement, this brings back eerie reminders of earlier advertisements that involved trade in human flesh (the reference point being the slave trade) and suggests that "we are fast returning to a world where persons carry a price tag, and where the cash value of some persons . . . is far greater than that of others."[14]

More sober voices, such as the scientist Doris T. Zallen, find themselves struggling to gain a hearing above the din of the rhetoric of enthusiasm. Having noted that the early promises of genetic intervention to forestall "serious health problems, such as sickle-cell anemia, cystic fibrosis, and Huntington disease," have thus far had only the most meager success, Zallen takes up a booming genetic enterprise that promises not merely the prevention of harm but the attainment of perfection through so-called genetic enhancement. One starts with a healthy person and then moves to perfect. Zallen calls this the "genetic equivalent of cosmetic surgery." The aim is to make people "taller, thinner, more athletic, or more attractive." She lists potential sources of harm, including reinforcement of "irrational societal prejudices." For instance, she continues, "what would happen to short people if genetic enhancement were available to increase one's height?"[15]

The "historical record is not encouraging," adds Zallen, noting the earlier eugenics movements with their hideous outcomes, most frighteningly in Nazi Germany but evident in this country as well where policies of involuntary sterilization of persons with mental retardation and other measures went forward apace. Many supporters of genetic enhancement will cry foul at this point, but where else can a quest for perfection lead except to elimination of the "imperfect" through improved and presumably more humane techniques that by now routinize aborting of infants deemed imperfect? For example: nearly 90 percent of detected Down syndrome pregnancies currently end in abortion. Medical pressure lies entirely in this direction; indeed, women report being accused of irresponsibility should they *choose*—and remember that "choice" is the regnant human good within this moral universe—to carry a Down syndrome child to term.

Calmer voices remind us that the scientific community at present has only the "vaguest understanding" of the details of genetic instruction—which is unsurprising, when one considers that each "single-celled conceptus immediately after fertilisation [*sic*]" involves a "100-trillion-times miniaturised information system," observes James Le Fanu.[16] Yet the enthusiasts who claim that the benefits of genetic manipulation are both unstoppable and entirely beneficial downplay any and all controversies, and they short-circuit any and all difficulties. In this way, they undercut (or attempt to) any and all "nonexpert" criticism in a manner that "effectively precludes others coming to an independent judgment about the validity of their claims."[17] The upshot is that it is very difficult to have the ethical and cultural discussion we require. Those who try to promote conversation and disciplined progress are tagged with the labels of "technophobes" or "bioluddites."

Despite this, there are a few critical straws in the cultural wind. In the 1997 film *Gattaca*, for example, the protagonist (played by Ethan Hawke) is born the "old-fashioned way" (a "faith-birth") to his parents, who had made love and taken their chances with whatever sort of offspring might eventuate. In this terrible new world, when a child is born, an immediate genetic profile is drawn. Our protagonist, Vincent, is a beautiful but, it turns out, genetically hapless child (on the standards of the barren world that is to be his lot) who enters life not amid awe and hopefulness but with misery and worry. His mother clutches the tiny newborn to her breast as his genetic quotient is coldly read off by the expert. "Cells tell all," the prophets of genetocism intone. Because of his genetic flaws—a result of his unregulated birth—young Vincent is not covered by insurance; he does not get to go to school past a certain age; and he is doomed to menial service. He is a "degenerate." Or, as the scanners immediately pronounce him, an "Invalid."

Gattaca's Vincent contrives a way to game the system as he yearns to go on a one-year manned mission to some truly far-out planet. Only "Valids"—genetically correct human beings—are eligible for such elite opportunities. So Vincent pays off a "Valid" for the Valid's urine, blood, saliva, and fingerprints and begins an arduous, elaborate ruse. For this is a world in which any tissue scraping—a single eyelash, a single bit of skin sloughing—might betray you. Why would a Valid sell his bodily fluids and properties? Because the Valid is now "useless," a cripple, having been paralyzed in a car accident. Indeed, his life is so useless on society's standards (and that notion, in turn, has been thoroughly internalized) that at the film's conclusion, and after having stored sufficient urine and blood so that Vincent can fool the system for years to come, the crippled Valid manages to ease himself into a blazing furnace to incinerate himself—life not being worth living any longer, not for one who cannot use his legs.

As for Vincent, despite some very tense moments, life is as good as it is ever going to get by film's end: He has made love to Uma Thurman and he has faked his way (with the connivance of a sympathetic security officer) onto the mission to the really distant planet of which he has dreamt since childhood, despite his genetically flawed condition. *Gattaca* is a bleak film. The only resistance Vincent can come up with is faking it. He has no language of protest and no ethical distance available to him. This is just the world as he and others know it and presumably will always know it. Uma Thurman's intimacy with an Invalid is as close to resistance as she can get. There are no alternative points of reference or resistance.[18]

Of course, we are not in the *Gattaca* nightmare yet. But are we drawing uncomfortably close? There are those who believe so, including the mother of a Down syndrome child who wrote the author of a column about genetic engineering in *The New Republic*.[19] That piece reflects on what our quest for bodily perfection might mean over the long run for the developmentally different. This mother, whose child died of a critical illness in his third year, wrote that she and her husband were enormously grateful to have had "the joyous privilege of parenting a child with Down Syndrome." She explains:

> Tommy's [not his real name] birth truly transformed our lives in ways that we will cherish forever. . . . But how could we have known in advance that we indeed possessed the fortitude to parent a child with special needs? And who would have told us of the rich rewards? . . . The function of prenatal tests, despite protestations to the contrary, is to provide parents the information necessary to assure that all pregnancies brought to term are "normal." I worry not only about the encouragement given to eliminating a "whole category of persons" (the point you make), but also about the prospects for respect and treatment of children who come to be brain-damaged either through unexpected birth traumas or later accidents. And what about the pressures to which parents like myself will be subject? How could you "choose" to burden society in this way?[20]

Expanding Choice and Diminishing Humanity

In the name of expanding choice, then, we are narrowing our definition of humanity and, along the way, a felt responsibility to create welcoming environments for all children, for we can simply declare that they chose to have an "abnormal" child and now they must pay the consequences. This declaration, if it is generalized, takes us, as individuals and a society, off the hook for the purpose of social care and concern for all persons, including those with bodies and minds that are not "normal." This trend stitches together a cluster of views under the rubric of expanding choice, enhancing control, and extending freedom. The end result is diminution of the sphere of the "unchosen" and expansion of the reign of "control-over." Rather than viewing children who are not "normal" in their development as a type of child who occurs from time to time among us and who, in common with all children, makes a claim on our tenderest affections and most fundamental obligations, we see such children as beset by a "fixable" condition: There must be a cure. The cure, for the most part, is to gain sufficient knowledge (or at least to claim to have such knowledge) that one can predict the outcome of a pregnancy and move immediately to prevent a "wrongful" birth in the first place. The fact that "curing" Down syndrome means one eliminates entirely one type of human being is no barrier to this effort. People living with Down syndrome must simply cope with the knowledge that our culture's dominant view is that it would be better were no more of their "kind" to appear among us.

In *The Future of the Disabled in Liberal Society: An Ethical Analysis*,[21] the philosopher Hans S. Reinders, a professor of ethics at the Vrije University in Amsterdam, argues that, despite public policy efforts to ensure equal opportunity and access for all, liberal society (including our own) cannot sustain equal regard for persons with disabilities. This is especially true if the disabilities in question are "mental." The liberal presupposition that privileges "choice" as the primary category in public life and the apogee of human aspiration, paired with modern technologies of reproductive and genetic engineering, dictates that it would be far better if human persons who are incapable of choosing on this model were not to appear among us.

So strong is the prejudice in this direction that we simply assume that hypothetical unborn children with cognitive disabilities would, if they could, choose not to be born. Reinders argues that the regnant view among liberal philosophers is that human beings with mental retardation may be regarded as members of the human species but they do not have full moral standing in the secular community. Because they lack such standing, the barriers to eliminating such persons will slowly but surely wither away. Given the religious derivation of so much of our ethical thinking, barriers simply to killing persons with disabilities remain. But such barriers, Reinders argues, are under continuous pressure from "secular morality" and are likely to be bulldozed out of the way by the potent machine of biotechnology backed up by medical authority. So it is not at all irrational for those with mental disabilities and their families to worry about the future. The proliferation of genetic testing, Reinders concludes, will most certainly have discriminatory effects because it puts everything under the domain of "choice," and parents of children with "special needs" become guilty of irresponsible behavior in "choosing" to bear such children and in burdening society in this way.

Increasingly, society expects, and even insists, that parents must—for this is the direction "choice" takes at present—rid themselves of "wrongful life" to forestall "wrongful births" that would burden them and, even more important, the wider society. Women repeatedly tell stories of the pressure from their medical caregivers to abort should a sonogram reveal something suspicious. The current abortion regime often embodies in practice a burden for women who are told that they alone have the power to choose whether or not to have a child and that they alone are expected to bear the consequences if they do not choose to do so. The growing conviction that children with disabilities ought never to be born and that the prospective parents of such children ought always to abort undermines the felt skein of care and responsibility for all children.

This is at least a reasonable worry, especially when the machinery of technology now surrounding childbirth turns every pregnancy into what was once called a "crisis pregnancy." Health maintenance organizations (HMOs) are now standardizing prenatal testing and genetic screening procedures that were once called upon only when couples had a history of difficulties. The point of all this is to initiate a process—should a sign even be hinted at—"of cajoling and pressuring that terminates in an abortion," observes the family life minister Jeannie Hannemann. She finds:

> A culture shift taking place moving away from supporting families with special-needs children toward resenting such families as creating a "burden" on society. . . . HMOs refusing treatment to special needs children,

arguing their mental or physical problem represented a "preexisting condition" because their parents elected not to abort them after prenatal screening indicated a problem.[22]

Appreciating the Complexity of Human Embodiment

The heart of the matter lies in a loss of appreciation of the complex nature of human embodiment. The social imagination—shaped by the dominant scientific voices in genetic engineering, technology, and "enhancement"—declares the body to be a construction, something we can invent. We are loathe to grant the status of *givenness* to any aspect of ourselves, despite the fact that human babies are wriggling, complex little bodies preprogrammed with all sorts of delicately calibrated reactions to the human relationships that "nature" presumes will be the matrix of child nurture. If we think of bodies concretely in this way, we are propelled to ask ourselves questions about the world little human bodies enter: Is it welcoming, warm, and responsive? But if we tilt in the biotechnological constructivist direction, one in which the body is so much raw material to be worked upon and worked over, the surround in which bodies are situated fades as The Body gets enshrined as a kind of messianic project.

In this latter scenario, the body we currently inhabit becomes the *imperfect body* subject to chance and the vagaries of life, including illness and aging. This body is our foe. The *future-perfect body* extolled in manifestos, promised by enthusiasts, and embraced by many ordinary citizens is a gleaming fabrication. For soon, we are promised, we will have found a way around the fact that what our foremothers and forefathers took for granted—that the body must weaken and falter and one day pass from life to death—will soon be a relic of a bygone era. The future-perfect body will not be permitted to falter. Yes, the body may grow older in a strictly chronological sense, but why should we age? So we devise multiple strategies to fend off aging, even as we represent aging bodies as those of teenagers with gleaming gray hair. A January 30, 2000, *New York Times Magazine* lead article by Stephen Hall, titled "The Recycled Generation," extolled the "promise of an endless supply of new body parts" via stem cell research, although that research is now "bogged down in abortion politics and corporate rivalries."[23] One of the entrepreneurs who stands to make even more millions of dollars in what the article calls the "scientific chase" for "the mother of all cells—the embryonic stem cell" bemoans the fact that the rush forward is being slowed down by a terrible problem, namely, the "knee-jerk reaction" on the part of many people to "words like 'fetal' and 'embryo.' "[24]

Disregarding Critiques and Concerns

The image that comes bounding out of Hall's *New York Times Magazine* piece is that genetic innovators who face opposition from religious and superstitious people, who go "completely irrational" when they hear certain words, are nonetheless fearlessly forging ahead in the teeth of sustained opposition—thus reversing the actual situation in which critics are the ones fighting a rear-guard battle against powerful, monied, influential cultural forces who, in line with the story our culture likes to tell about itself, represent "progress" and a better future.[25] The upshot is that rather than approaching matters of life, death, and health with humility, knowing that we cannot cure the human condition, we seek cures on the assumption that the more we control the better. During work on this chapter, word came that a human embryo had been cloned—probably a fraud, in this particular case, but the most interesting thing is the media reaction to this announcement. Television commentary resounded with the hyperbole that this will make possible, in the future, a near endless supply of body parts that can be harvested to indefinitely prolong human life. Thus, even before a grown clone appears—and let us pray this does not happen—the clone is reduced to property to be harvested for the benefit of others.

The underlying presupposition is, of course, that nothing is good in itself, including embodied existence. Thus, it becomes easier to be rather casual about devising and implementing strategies aimed at the selective weeding out or the destruction of the bodies of those considered imperfect or abnormal or even the bodies of the "perfect" if that human entity is cloned. "A science fiction dystopian mentality" is blamed for any questions about whether the path we are racing down might not turn old age itself into a pathology and usher in cultural "encouragement" for the "unproductive" elderly to permit themselves to be euthanized because they are extra mouths to feed and a nuisance to just about everybody.

It is difficult to overstate just how widely accepted the technocratic view is and how overwhelmingly we, as a culture, are acquiescing in its premises. In an August 8, 1996, *Times Literary Supplement* review of four new books on the genetic revolution, the reviewer matter-of-factly opined that "we must inevitably start to choose our descendants," adding that we do this now in "permitting or preventing the birth of our own children according to their medical prognosis, thus selecting the lives to come."[26] This point is considered so unexceptionable that it is not argued for but rather merely assumed as a given. As long as society does not cramp our freedom of action, we will stay on the road of progress and exercise sovereign choice over birth by consigning to death those with a less than stellar potential for a life not "marred by an excess of pain or disability."[27]

In 1969, the molecular biologist Robert L. Sinsheimer, who originated the idea of sequencing the human genome, called for a "new eugenics"—a phraseology most try to avoid given the association of the "e-word" with the biopolitical ideology of mid-twentieth-century National Socialism. In Sinsheimer's words, "The new eugenics would permit in principle the conversion of all the unfit to the highest genetic level."[28] The popular press tracks our society's responses in widespread adoption of prenatal screening, now regarded as routine—so much so that prospective parents who decline this panoply of procedures are treated as irresponsible—we see at work the presumption that life should be wiped clean of any and all imperfection, inconvenience, and risk.[29] Creation itself must be put right.

The *New York Times* of December 3, 1997, alerted us to this fact in an essay, "On Cloning Humans, 'Never' Turns Swiftly into 'Why Not,'" by its science editor, Gina Kolata (and the "why not" has now been done, or so is the claim). Kolata points out that in the immediate aftermath of Dolly, the cloned sheep who stared out at us from the covers of so many newspapers and magazines, much consternation and rumbling arose.[30] But opposition dissipated quickly, she continues, with fertility centers soon conducting "experiments with human eggs that lay the groundwork for cloning. . . . Moreover, the Federal Government is supporting new research on the cloning of monkeys, encouraging scientists to perfect techniques that could easily be transferred to humans."[31]

A presidential ethics commission, under Bill Clinton, may have recommended a "limited ban on cloning humans," but after all, "it is an American tradition to allow people the freedom to reproduce in any way they like"—according to Laurence H. Tribe of Harvard Law School.[32] This claim is simply a distortion of the historic and legal record. In common with any society of which we have any knowledge, past or present, American society has built into its interstices a variety of limitations on "reproductive freedom." But the view that "freedom" means doing things in "any way one likes" now prevails as a cultural desideratum.[33] It is therefore unsurprising that Kolata's *New York Times* article describes a "slow acceptance" of the idea of cloning in the scientific community that took but six months to go from shock and queasiness to acquiescence and widespread approval. Kolata concludes that "some experts said the real question was not whether cloning is ethical but whether it is legal." And one doctor is quoted in these words:

> The fact is that, in America, cloning may be bad but telling people how they should reproduce is worse . . . In the end . . . America is not ruled by ethics. It is ruled by law.[34]

The implication of this view is that no ethical norm, standard, commitment, or insight can or should be brought to bear whether to criticize, to caution against, or to checkmate statutory laws should they be unjust or unwise. The point is that with each new development presented to us in the name of a radical and benign extension of human freedom and powers, we pave additional miles on the fast track toward eradication of any real integrity to the category of "the human." Debate and discourse about such matters in the public square has turned into a routine in which a few religious spokespersons are brought on board to fret a bit and everything marches on.[35]

That the prospect of human cloning—or at least one dominant theme or strand—is fueled by narcissistic fantasies of radical sameness, that it represents fear of the different and the unpredictable, that it speaks to a yearning for a world of guaranteed self-replication matters not—indeed, such concerns are rarely named save by those speaking from the point of view of theological anthropology. One such is the Pontifical Academy for Life, in a statement on human cloning issued June 25, 1997:

> Human cloning belongs to the eugenics project and is thus subject to all the ethical and juridical observations that have amply condemned it. As Hans Jonas has already written, it is "both in method the most despotic and in aim the most slavish form of genetic manipulation; its objective is not an arbitrary modification of the hereditary material but precisely its equally arbitrary fixation in contrast to the dominant strategy of nature."[36]

The popular press tracks the fascination with cloning.[37] Dreams of wholesale self-possession grounded in attaining full control over human "reproductive material" lie at the heart of the eugenics project, despite the risk of damaging biogenetic uniformity, because much of the basic genetic information that goes into the creation of a child from two parents emerges as a result of sexual reproduction.

Following "the Wisdom of Repugnance"

What, then, about embarking on an experimental course likely to result in flawed "products?"[38] It is convenient to forget that it took nearly 300 failed attempts before Dolly the sheep was cloned successfully. As Leon R. Kass has noted, the image of failed human clones leads the soul to shudder. Abandoning what he calls "the wisdom of repugnance," we embark on a path that constitutes a violation of a very fundamental sort. He calls upon us to pay close attention to what we find "offensive," "repulsive," or "distasteful," for such reactions often point to deeper realities.[39] He explains why:

> In this age in which everything is held to be permissible so long as it is freely done, in which our given human nature no longer commands respect, in which our bodies are regarded as mere instruments of our autonomous rational wills, repugnance may be the only voice left that

speaks up to defend the central core of our humanity. Shallow are the souls that have forgotten how to shudder.[40]

Kass is not arguing that repugnance is the end of the matter but instead that it is a beginning. Those philosophies that see in such reactions only the churnings of irrational emotion misunderstand the nature of human emotions. Our emotional reactions are complex, laced through and through with thought. The point is to bring forward such reactions and submit them to reflection.

Would we really want to live in a world in which the sight of piles of anonymous corpses elicited no strong revulsion, or a world in which the sight of a human being's body pierced through in dozens of places and riddled with pieces of metal was something we simply took for granted? The reaction to the first clearly gestures toward powerful condemnation of those responsible for creating those mountains of corpses, and anguish and pity for the tortured and murdered and their families. In the case of the metal-pierced-body, we may decide it is a matter of little import and yet ask ourselves why mutilation of the body that goes much beyond the decorative is now so popular? Does this tell us anything about how we think about our bodies?[41] And so on.

The "technical, liberal, and meliorist approaches all ignore the deeper anthropological, social, and, indeed, ontological meanings of bringing forth new life," asserts Kass. "To this more fitting and profound point of view, cloning shows itself to be a major alteration, indeed, a major violation, of our given nature as embodied, gendered and engendering beings—and of the social relations built on this natural ground."[42] The upshot is that critical interpreters cede the ground too readily to those who want to move full steam ahead when in fact it should work the other way around. "The burden of moral argument must fall entirely on those who want to declare the widespread repugnances of humankind to be mere timidity or superstition."[43] Too many theologians, philosophers, and cultural critics have become reticent about defending insights drawn from the riches of the Western tradition.

As a result, Kass argues, we do the following things: We enter a world in which unethical experiments "upon the resulting child-to-be" are conducted; we deprive a cloned entity of a "distinctive identity not only because he will be in genotype and appearance to another human being, but, in this case, because he may also be twin to the person who is his 'father' or 'mother'—if one can still call them that"; we deliberately plan situations that we know—the empirical evidence is incontrovertible—are not optimal arenas for the rearing of children, namely, family fragments that deny relationality or shrink it. Finally, he suggests:

> [We] enshrine and aggravate a profound and mischievous misunderstanding of the meaning of having children and of the parent–child relationship. . . . The child is given a genotype that has already lived. . . . Cloning is inherently despotic, for it seeks to make one's children . . . after one's own image . . . and their future according to one's will.[44]

The many warnings embedded in the Western tradition, from its antique (pre-Christian) forms through Judaism and Christianity, seem now to lack the power to stay the hand of a "scientized" anthropocentrism that distorts the meaning of human freedom.[45]

Understanding the "Natural": A Christian Anthropology

Within the Jewish and Christian traditions, a burden borne by human beings after the Fall lies in discerning what is natural or given, presuming that what is encoded into the very nature of things affords a standard, accessible to human reason, by which we can assess critically the claims and forces at work in our cultural time and place. (This is not the only available standard, of course, but it was long believed an important feature of a whole complex of views.) The great moral teachers, until relatively recently, believed that "nature" and "the natural" could serve as a standard. Within Christian theological anthropology, human beings are corporeal beings—ensouled bodies—made in the image of their Creator.

John Paul II

According to the late Pope John Paul II, this account of our natures, including the ontological equality of male and female as corporeal beings, is "free from any trace whatsoever of subjectivism." As he explains, "it contains only the objective facts and defines the objective reality, both when it speaks of man's creation, male and female, in the image of God, and when it adds a little later the words of the first blessing: 'Be fruitful and multiply and fill the earth; subdue it and have dominion over it (Gen. 1:28)."[46] Dominion here—it is clear from the overall exegesis—is understood as a form of stewardship, not domination.

John Paul's account of Genesis is presaged in his prepapal writings (as Karol Wojtyla). For example, in a series of spiritual exercises presented to Pope Paul VI, the papal household, and to the cardinals and bishops of the Roman Curia during a Lenten Retreat in March 1976, Karol Cardinal Wojtyla argued:

> One cannot understand either Sartre or Marx without having first read and pondered very deeply the first three chapters of Genesis. These are the key to understanding the world of today, both its roots and its extremely radical—and therefore dramatic—affirmations and denials."[47]

Teaching about human origins, human beginnings, in this way offers "an articulation of the way things are by virtue of the relation they have with their creator."[48] Denying that relationship, we too easily fall into subjectivism, into a world of rootless wills.

Dietrich Bonhoeffer

With this sense of the natural, Dietrich Bonhoeffer would agree. In his discussion of "The Natural" and "The Right to Bodily Life" in his *Ethics*—left incomplete when he was arrested by the Gestapo and then hanged for his part in the plot to assassinate Adolf Hitler—Bonhoeffer observes that "the natural" fell out of favor in Protestant ethics and became the almost exclusive preserve of Catholic thought. For Protestants, the natural was so corrupted through the Fall that it could never serve as a standard of assessment. The fear was that grace would thereby be diminished. Bonhoeffer aims to resurrect "the natural," insisting that human beings still have access to the natural but only "on the basis of the Gospel."[49]

What is Bonhoeffer suggesting? It means that a harsh antithesis between nature and grace should be abandoned in favor of a more complex view. The Gospel helps to uncover this complexity in this way: The natural is "directed towards the coming of Christ," he tells us. "The unnatural is that which, after the Fall, closes its doors against the coming of Christ."[50] How so? Because absent affirmation of the Incarnation, which redeems the natural world and penultimate, the natural must necessarily be an inadequate, even corrupt, standard. Any view of nature and the natural that desacralizes the penultimate by presuming that the immanent exhausts lived life actually demeans the human person. The unnatural "is exposed once and for all as destruction of the penultimate."[51] The natural, correctly understood, is the very form of life. Assault against that form is an attack on reason itself in favor of a destructive, untrammeled will. "If life detaches itself from this form," he writes, "if it seeks to break free and to assert itself in isolation from this form, if it is unwilling to allow itself to be served by the form of the natural, then it destroys itself to the very roots." And so it is that "life which posits itself as an absolute, as an end in itself, is its own destroyer."[52]

In his move to redeem the concept of the natural, Bonhoeffer argues that human beings enjoy a "relative freedom" in natural life. But there are "true and . . . mistaken" uses of this freedom and these mark the difference between the natural and the unnatural. Thus it follows: Destruction of the natural means destruction of life. . . . The unnatural is the enemy of life. It violates our natures to approach life from a false "vitalism" or excessive idealism on the one hand or on the other from an equally false "mechanization" and lassitude that shows "despair towards natural life" and manifests "a certain hostility to life, tiredness of life and incapacity for life." Our right to bodily life is a natural, not an invented, right and the basis of all other rights, given that Christians repudiate the view that the body is simply a prison for the immortal soul.

Harming the body harms the self at its depth. "Bodilyness and human life belong inseparably together," in Bonhoeffer's words. Our bodies are ends in themselves. This has "very far-reaching consequences for the Christian appraisal of all the problems that have to do with the life of the body, housing, food, clothing, recreation, play and sex." We can *use* our bodies and the bodies of others well or ill. The most striking and radical excision of the integrity and right of natural life is "arbitrary killing," the deliberate destruction of "innocent life." Bonhoeffer specifically notes as examples abortion, killing defenseless prisoners or wounded men, and destroying lives we do not find worth living—a clear reference to Nazi euthanasia and genocidal policies toward the ill, the infirm, and all persons with handicaps.[53] "The right to live is a matter of the essence" and not of any socially imposed or constructed values. Even "the most wretched life" is "worth living before God." Other violations of the liberty of the body include physical torture, arbitrary seizure and enslavement (American slavery is here referenced), deportations, and separation of persons from home and family—the full panoply of horrors the twentieth century dished up in superabundance. This fragment by Bonhoeffer on the natural is powerfully suggestive and worth pondering as an alternative to those cultural dictates that declare any appeal to nature or the natural as a standard illegitimate.

Bonhoeffer rather uncannily anticipates the ethics of the present moment that advances a crude standard of utility by which to measure whether something should go forward or not—whether in technology, science, education, any other field of endeavor—and this under a version of "the greatest good for the greatest number." He criticizes such a view for turning the individual into a "means to an end in the service of the community," citing the results:

> The happiness of the community takes precedence over the natural right of the individual. This means in principle the proclamation of social eudemonism and the curtailment of all the rights of individuals.[54]

In Bonhoeffer's era, this meant the triumph of the Nazi Reich over the human person and his or her most fundamental right, that to natural life itself. We have our own form of social eudemonism: It would be better for the unfit never to be born, for society would yield a net benefit from this policy. It would be better for the unproductive elderly to die, for society would similarly yield a net benefit. This eudemonism is even taken down to the family level: It would be better for the family not to have to care for a disabled newborn. The greater good of the family would be served thereby, as would the greater good of the whole. Bonhoeffer is right: This social eudemonism "allies itself with a blind voluntarism," and it involves an "inconceivable overestimation of the power of the will in its encounter with the reality of natural life itself."[55]

It goes without saying that more work would need to be done in order to redeem the categories of "nature" and "the natural," but it is sufficient here to point out that our present circumstances resist the possibility of this conceptual and ethical possibility even as the need for some such standard becomes ever more urgent. We need powerful and coherent categories and analysis that challenge cultural projects denying finitude, promising a technocratic agenda that ushers in almost total human control over all of the natural world, including those natures we call human, pushing toward an ideal of sameness through genetic manipulation and self-replication via cloning, and continuing with the process of excision of bodies deemed unworthy to appear among us and to share our world. Perfection requires manipulation and elimination—a kind of purificationist imperative is at work here, aiming to weed out the flawed and recognizing only the perfect and fit. Wrapped up in a quest for control, immersed in the images and rhetoric of choice and self-possession, we will find it more and more difficult to ask the right sorts of questions as we slowly but surely lose rich languages of opposition, as embodied in Christian theological anthropology.

Conclusion

Christian anthropology affirms that one is born "*in communio*"—that is, "in communion." None of us are single, solitary, atomistic individuals. We are born into a world of others (and the Other) and are therefore bound to them and them to us. Furthermore, scripture uses the powerful metaphor of "the body" to describe those who share a common union with Christ. Martin Luther, in his "Gospel for the Early Christmas Service," describes the church not as the building in which believers meet but as "the assembly of people who believe in Christ." That means that "with this church one should be connected and see how the people believe, live, and teach."[56]

Recovering an embodied community of wisdom should be the aim of much of our efforts together. Because Christians participate in community in virtue of their union with Christ, community is not, said Luther, "an ideal we have to realize" but a reality in which we share. "Life together," to borrow Bonhoeffer's famous title, requires that we are stirred to love of neighbor.[57] Real community will involve embodied compassion. Real community will manifest itself in acceptance of others as embodied imagers of God, whether they are naturally weak or naturally strong, whether fully abled or less fully abled. After all, we share in tactile, relational, fleshly, and concrete ways, and the Incarnate God sanctified the tactile, relational, and fleshly by coming to us in a human body.

Christian communities are to embody not only God's compassion but also God's wisdom. Together, Christians enjoy the indwelling presence of the Holy Spirit, who is able to lead them into truth. Wisdom-formed communities provide ballast in a world where the quest for control of one's own body and one's own destiny has become all consuming. Others should be able to observe in a wisdom-formed community "how people believe, live, and teach."

The challenge for the church, then, is to embody such a community. Without a vibrant, living, Christian community, the rest of the world will be unable to witness a way of life different from that of the dominant culture. The values and practices that have shaped a way of life obsessed by choice, consent, and control must be answered with values and practices that shape a community marked by love, care, sacrifice, and trust.

Wisdom-formed communities are not necessarily nontechnological communities. Instead, they are communities in which social practices are informed not only by the values embodied by the community but also by the experience of community. That is to say, Christian communities must measure technologies, including biotechnologies, by the ways these technologies either diminish our shared humanity or contribute to our life together. For instance, electronic technologies like e-mail, text-messaging, and synchronous chat must not become a substitute for embodied community experiences but rather a facilitator of those experiences. Otherwise, those technologies become a means of subtly demonstrating that we "believe, live, and teach" that the Incarnation is unimportant.

What follows in the next chapter is an examination of a number of technologies that promise to enhance control of one's own individual goals. Those technologies and their impetus must be subjected to scrutiny. Do they, in fact, enhance our embodied community, our life together, or do they further alienate, distance, and separate us from the community? Do they contribute to the common good of the community, or do they only serve the instrumental ends of the one who employs them? These are the kinds of questions that must be answered before we embrace new technologies. Even more urgent is it for the church to answer these questions, for in embracing new ways of living, we alter the way we testify to what we "believe, live, and teach."

CHAPTER SIX

Biotechnology, Human Enhancement, and the Ends of Medicine

Technology opens doors, it does not compel man to enter.

—Lynn White Jr., *Medieval Technology and Social Change*

ONE of humanity's persistent dreams has been to seize control of nature's laws and eradicate the fragility and finitude of human life. For most of human history, this elusive hope resided only in the literary imagination. Writers fashioned a multitude of utopias free of strife, aging, disease, and death.[1] Others just as zealously exposed the nightmares and follies of human attempts to outdo the Creator.[2] From the sixteenth and seventeenth centuries on, however, science and technology began to place the means to control nature in human hands, and the elusive dream took on the air of reality.

First to come under control was the physical world, culminating in the twentieth century in the conquests of space and atomic energy. In the last half of that same century, control extended to biology, to the secrets of human life itself. Twenty-first-century biotechnology in the form of molecular biology, genetics, nanotechnology, cybernetics, and psychopharmacology promises to eradicate disease, to go beyond mere therapy to enhance the quality of every facet of life.[3] Beyond all this is the lure of transforming human nature itself to free it even from the vagaries of natural selection, evolution, and ecology.[4]

As a result, humanity's recurrent dream has been revitalized. Today's scientists, philosophers, and pundits see modern humanity freed of gods and God. Their unbridled enthusiasm scoffs at Ecclesiastes' mordant wisdom. They dismiss that author's pessimism about the vanity of all things under the sun because there is indeed something new under the sun—biotechnology. Humanity can become its own redeemer, as one of medicine's sagest practitioners, the Canadian physician Sir William Osler, predicted in a moment of joyous adulation at early-twentieth-century medicine's prowess.[5] Osler scarcely could have envisioned that his beloved art would go beyond the eradication of disease to the satisfaction of every human desire, not just for health but for physical, physiological, and emotional perfection.

To use Lynn White's metaphor, biotechnology has indeed opened a wide, new, and confusing array of doors.[6] Today we must decide which of those doors to enter, which to explore tentatively, and which to keep tightly shut. More than anything else, we must control our power to control who, and what, we are. Otherwise, we are in danger of becoming victims of our own ingenuity, in which we make our utopias into dystopias.

Sadly, however, there is no historical evidence that technology can be limited by moral constraint, or that what starts as legitimate treatment of disease will not be used beyond therapy.[7] To exert moral constraint requires grappling with what it is to be human. This is the crucial, first-order philosophical and theological question that creates the deepest fault lines in contemporary culture. The President's Council on Bioethics has clearly recognized this fact.[8] The council's clarity in defining the deeper issues and its call for a "richer bioethics" are essential first steps. But this is just the beginning, for it is beyond this first step that the ethical quandary begins.

Against this background, we examine the narrower question of medicine's relationship with biotechnology. To what extent should medicine and physicians become the vehicles for individual and societal access to technobiology's promised benefits? In the realm of disease treatment, there is little question that physicians are the logical and necessary agents. But what about the "enhancement" of individual and social life, or the promises of perfection of human nature itself, beyond therapy?

To what extent should medicine be biotechnology's servant? Should medicine be redesigned to accommodate biotechnology? Should a new profession be created for this purpose? Are the aims of biotechnology "good" for patients or for humans as humans? How would being a Christian physician influence the responses to these questions?

This chapter focuses on the relationships between biotechnology, enhancement, and the ends of medicine from both secular and Christian perspectives. We begin with the difficulties of attempting to define the key terms: *health*, *disease*, *illness*, and *sickness*. Given the variability of meanings, operating definitions are offered, especially for the newest term: *enhancement*. These operating definitions are then related to the ends of medicine, conceptually, and then at three points where biotechnology intersects with medicine: (1) in disease treatment, (2) in "enhancement" beyond therapy, and (3) in reshaping human society and human nature. Each intersection is examined from the ethical, philosophical, and religious points of view.

The Key Concepts: Conceptual and Historical Difficulties

If the ends of medicine are to serve as boundary conditions for inclusion or exclusion of any form of biotechnology within or beyond clinical medicine, the vexations of defining certain concepts—health, disease, illness, healing, medicine itself, and the new term enhancement—must be confronted. The literature, both contemporary and historical, is vast. The concepts and their associated issues are well represented in two excellent collections of opinions, *Concepts of Health and Disease: Interdisciplinary Perspectives*, edited by Arthur L. Caplan, Tristram Engelhardt Jr., and James J. McCartney; and *Health, Disease, and Illness: Concepts in Medicine*, edited by Arthur L. Caplan, James J. McCartney, and Dominic A. Sisti.[9]

Health

"Health" is the ultimate end of medicine as medicine for individuals and for society. It is perhaps the most difficult of the key concepts to define. The word *health* comes from the Old English word meaning "wholeness." Leon R. Kass takes this to be a static concept, in contrast with the Greek idea of *hygei* and *euexi,* "a well way of living" and "well habitness." Kass prefers the Greek sense of health as "the well-functioning of the organism as a whole . . . an activity of the body in accordance with its specific excellences."[10]

In the ancient world, health had a variety of meanings with subtle differences between and among them. For the Pythagoreans, for example, health was a matter of balance with respect to food, activity, and environmental exposure—essentially the avoidance of immoderation of the appetites.[11] For the Hippocratic physicians, health also was a balance, but between the four humors.[12] Sextus Empiricus ranked health as the greatest good, while others put virtue ahead of health as a human good. Galen had a more down-to-earth definition, calling health "a state in which we neither suffer pain nor are hindered in the functions of daily life."[13]

In the modern era, the definition of health and its relationship with disease has become more complicated. This is the result of the dichotomy that has arisen between the value-free scientific concept of disease and the large number of value-laden socially and culturally determined definitions.[14] In the Caplan, Engelhardt, and McCartney collection of health concepts, we find health defined in different ways, depending on its relationship with opposing views of disease. Most of these emphasize value-laden concepts, such as health as a regulative ideal (in an essay by Engelhardt), optimum capacity for performance of the role to which one is socialized (Parsons), and conformity with prevailing cultural ideals (King; Fabrega).[15]

The most expansive definition of health—also defined in relation to the absence of disease—comes from the World Health Organization: "a state of complete physical, mental, and social well-being . . . a fundamental right of every human being."[16] It includes mental health, perhaps the most difficult of the concepts to define, as is "well being."[17] The World Health Organization's definition weakens any attempt to set boundaries for the perimeters of medicine or medical practice. To a certain extent, this would also be true of any of the most value-oriented definitions of health.

Disease

The definition of disease is as fraught with difficulty as the definition of health.[18] Like health, it can be defined objectively and scientifically, as Boorse does, or as the negation of some value-laden states defined culturally and socially.[19] Recently, Worrall and Worrall have suggested that all attempts to define the concept of disease are inadequate and that the effort should be abandoned.[20] Only specific diseases exist, each with its own defining signs, symptoms, and so forth. Though this analysis has some validity, it does not help much, because it seems to beg the question. After all, by what criteria do we determine which sets of signs and systems are specific diseases and which are not?

This is not the place to attempt a resolution of the epistemological problem, critical as it may be. Sooner or later, the discussion must settle on some operating definitions. Their validity will be revealed by the cogency of the conclusion to which they tend. With these caveats, we offer some definitions based on common usage and then link these usages to the ends of medicine.

Health is taken to be both a subjective and an objective state. Subjective health refers to that state in which a person feels that one's body, mind, and psyche, though not necessarily perfectly functioning, permit pursuing most of the things one wishes to pursue without undue discomfort. This is a paraphrase of Galen's simple definition given above.[21] Health, in this view, is a relative state in which each person strikes a balance between whatever limitations genetics, environment, or experience have imposed and one's aspirations.

One may feel "healthy" in this sense, even though one has a known disease or disability. The healthy person feels "well" so long as he, or she, can pursue goals other than to focus his entire attention on his bodily or mental apparatus. He becomes sick when his body becomes the focus of his life and an impediment to its pursuit. Subjective health is thus a highly personalized state of satisfaction or dissatisfaction determined by the patient himself.

Objective health, conversely, is a state in which no demonstrable physiological, psychological, or biochemical aberration from some objective standard of normality is present. In contrast with subjective health, objective health is determined by experts—persons other than the patient himself. Were we to examine every person, almost regardless of age and with the detail required, some quiescent or nascent pathological process would reveal itself. Few, if any, humans would qualify as objectively healthy.

Objective health is subject to change at any moment at the molecular, cellular, or organ tissue level. It may or may not manifest itself as a symptom or sign. A person may feel subjectively "healthy" in the presence of covert or even manifest disease, or one may have no objective disease and yet feel unhealthy. One may feel "unhealthy" even if without demonstrable disease if, for a multitude of reasons, one does not possess some attribute or attributes of body, mind, or character that one feels deprived of unfairly. This is often the case in the affluent, developed world of today.[22]

Disease is an objectively discernible abnormal state anatomically, physiologically, biochemically, or psychosocially. Disease is a state deviating by some order of magnitude, for example 2 standard deviations from the mean for a person's cohort group. Murphy has shown how confusing even the notion of normality can be.[23] Nonetheless, the concept is unavoidable in clinical and common parlance and in clinical decisions.

Disease, like health, is a continuum on which the physician must place a point that demarcates normal from abnormal. At one end of the continuum, the abnormality is so obvious that the untrained eye can readily detect it, for example, a neglected fungating carcinoma of the breast. At the other end, the demarcation is so subtle or nuanced that experts often differ, for example, deciding whether a mammographic image (lesion) can be "followed" or needs to be biopsied. To be sure, patient preferences, physicians' values, age, and resource availability may influence the decision, but a decision must be made. Is the finding "within the normal limits" or not?

A similar continuum exists for emotional and psychosocial disorders. Here, the objective determinants of the normal/abnormal decision are observation of behavior, psychometric testing, and, increasingly, biochemical studies, as well as sophisticated radiographic and electronic imaging. Cultural and social criteria may play a larger part, but some point still must be established on the continuum that separates normal from abnormal as a working, if not an ontological, distinction.

Illness and Sickness

Illness and sickness are subjective states in which the patient determines what is normal or abnormal. In both cases, a person perceives that "something" is wrong with bodily or mental functioning. There may or may not be symptoms that may be vague and diffuse. The patient experiences a state of unwellness, or a vague dissatisfaction with life's ordinary events. Often, nothing objectively abnormal is found, and most of the difficulty seems to be with one's life situation. Illness and sickness may accompany disease or be experienced in its absence. Not finding an objective sign does not mean that the patient's sense of illness is imaginary, only that it cannot be categorized. The person may or may not assume a "sick role," whether socially authenticated or self-designated. Feeling sick and illness are genuine sources of distress, unhappiness, and dysfunction. The resulting dissatisfaction, melancholia, or sense of deprivation offers fertile ground for some of the "enhancements" of quality of life that biotechnology promises.

The Ends of Medicine as Traditionally Defined

Before attempting to relate biotechnological enhancements to the ends of medicine, we must examine how the ends of medicine have traditionally been defined. We can then turn our attention to the meanings of enhancement to see if they fit properly within the domain of medical practice.

Aristotle begins his *Ethics* with the importance of "ends" and "the good" to the whole of ethics: "Every art and every inquiry, and similarly every action and pursuit, is thought to aim at some good; for this reason the good has been rightly declared to be that at which all things aim."[24] From here on, in this chapter the terms *end* and *good* will be used in the Aristotelian sense when applied to medicine.

In this view, the end of medicine is the good of the patient, and the end of social medicine is the good of society. Most of our focus is on clinical medicine. With respect to the larger issues of species transformation, the ethics of society is treated as well. Although the good of individual patients and the good of society are closely related, they are treated separately.

The end of the clinical encounter, from time immemorial, has been the treatment of disease and the relief of suffering. The end of medicine, in the words of the Hippocratic physician, is "to do away with the sufferings of the sick, to lessen the violence of their disease, and to refuse to treat those who are overmastered by their diseases since in such cases we are powerless."[25] Cure is not mentioned, for ancient medical knowledge rarely if ever could promise such an outcome. It is important to note that early in medicine's history, its limits were recognized and the notion of futility in certain cases was an acceptable reason for withholding treatment.

Over the intervening centuries, medical knowledge and techniques improved sufficiently so that cures became a more frequent possibility. The relief of suffering and amelioration of illness remained ends as well, especially when a cure was not possible. The prevention of disease and the positive cultivation of health were also among the ends of medicine from Hippocratic times.[26]

In contemporary medicine, two additional ends were added to the pristine ends of ancient medicine: the need for fusion of the technically correct with the morally good; and an awakened respect for the dignity of the patient, expressed as a recognition of a moral claim to participation in decisions affecting the patient. The good of the patient as the *telos* or end of medicine has become a complex, holistic notion. Healing now comes closer to its etymological sense of making "whole."[27]

We can separate the ends of medicine into proximate and ultimate. The proximate end is a fusion of the technical and the moral, that is, a technically correct and a morally right decision about the good of an individual patient. The ultimate goal is health, a concept that is difficult to define, as we have seen. Both ends require an up-to-date knowledge of the science of medicine and the clinical arts of diagnosis, prognostics, and therapy if healing is to be accomplished.

Today, the terms "goals" or "purposes" are often substituted for "ends." They are socially determined and defined by the mores of a particular community, time, and place. One may set many goals for medicine beyond healing that give medicine its essential character. Medicine in our time has, for example, been used for eugenic, genocidal, military, and other purposes. Ends, however, are not arbitrary. They derive from, and themselves define, the kind of activity that medicine is.[28]

For our purposes, the "ends" we are most concerned with are those of clinical medicine: the use of medical knowledge in treating and preventing disease (as defined above) in individual patients, as well as alleviating pain and suffering due to physical and/or emotional disease. Clinical medicine has as its particular end the good of individual persons seeking help through the use of medical knowledge and skill.

This is admittedly a disease-based approach, fully cognizant of the difficulties of such a position. It is, however, on balance preferable to those ideologies that define medicine more broadly in terms of whatever ends patients, physicians, or society choose to create.[29] Such definitions confuse purposes with ends and make physicians automatic servants of societal will or biotechnological possibility. They deprive society of whatever restraining influence physicians might exert over the more egregious abuses of knowledge or social whim a society might contrive.

Biotechnology, Medicine, and Enhancement

The actual and promised capabilities of biotechnology have given prominence to a possible new end of medicine, namely, *enhancement*. This term does not have the long history of the terms discussed above to designate the ends of medicine. Almost every present-day commentator underscores the difficulties, impossibility, or futility of any definition that seeks to distinguish enhancement from therapy.[30] Nonetheless, everyone sooner or later ends up using the term, because no viable substitute has yet appeared. This is so because some boundary between morally valid and invalid uses of biotechnology cannot be established without at least a working definition.

What most definitions of enhancement signify, in one way or another, is the notion of going "beyond" something—that is, beyond the generally accepted ends of medicine or personal well-being, or beyond the limitations of the human species itself. This is the sense of enhancement captured in the 2003 report *Beyond Therapy* by the President's Council on Bioethics.[31]

Eric Juengst puts it very succinctly: Enhancement covers "interventions designed to improve form or functioning beyond what is necessary to sustain or restore good health."[32] Most commentators seek some agreement on some standard by which moral constraint can be applied. This is true of even the more moderate effort by Brock to make public policy discussions "more consistent, nuanced and systematic."[33]

To be sure, where the boundaries are drawn, and for what purpose, usually depends on a particular perspective the author wishes to espouse. Much of Daniels' effort, for example, is an extension of his interests in just health care and insurance.[34] Others emphasize the broader issues of personal enhancement or improvement of human nature, or the theological dimensions. Suffice it to say that without a working definition, overt or covert, most discussions about enhancement would be difficult even to initiate.

Our operating definition of enhancement will be grounded in its general etymological meaning: "to increase, intensify, raise up, exalt, heighten, or magnify," are some of the usages in the *Oxford English Dictionary*. Each of these words carries the connotation of going "beyond" what exists at some moment, whether it is a certain state of affairs, a bodily function or trait, or a general limitation built into human nature. Enhancement is, as Fowler says, "a dangerous word for the unwary," but its use in some form seems inescapable.[35]

More specifically for this discussion, enhancement will signify an intervention that goes beyond the ends of medicine as described above. For medicine, the treatment/enhancement distinction cannot be avoided, because physicians will play a central role whenever medical knowledge is used both to regain health and to go beyond what is required to regain health. To be sure, specialists in the physical, biological, social sciences, and engineering are necessary if even the more modest promises of biotechnology are to be realized; they will provide the basic scientific and technical know-how from which biotechnological enhancements will emerge. But physicians are crucial in actually using this technology with individual human beings.

For one thing, physicians will be needed to design and conduct experimental protocols and clinical trials. They will be writing the prescriptions for "off-label" uses of enhancing intervention, because many of the agents in question are designed first for the treatment of specific diseases. The medicine Modafinil is a good example of the process. It was designed to treat narcolepsy, a serious sleep disorder, but already the great preponderance of prescriptions are written off-label, that is, not for narcolepsy but to alter sleep patterns and ward off sleepiness for any cause.[36] This is the route already taken by anabolic steroids, recombinant erythropoietin, and transfusions as performance enhancers in competitive sports.[37] In each instance, the initial therapeutic use is taken "beyond therapy."

Clearly, some physicians have already crossed the divide between treatment and enhancement, between medically indicated use and patient-desired abuse. Already physicians need to reflect on the ethical implications of their involvement in the uses of biotechnology. Our reflection centers on three uses: (1) the treatment of disease; (2) the satisfaction of patients' and nonpatients' desires for the enhancement of some bodily or mental trait, or some state of affairs they wish to perfect; and (3) (more distantly) the redesign of human nature and thus the enhancement of the species in the future.

Treatment of Disease

New treatments for both new and old diseases are the most promising uses of biotechnology. They most closely conform to the clinical and ethical ends of medicine. The list of target diseases is long. Devising treatments for them is a legitimate, and desirable, individual and social good. Here, the physician functions in the time-honored role as healer, with a moral obligation to stay informed and educated in using the new technologies. The ethical questions are related to the means by which these new treatments are developed and applied. Genetic manipulations, cybernetics, nanotechnology, and psychopharmacology are in themselves neither intrinsically good nor bad morally. However, procedures derived from the destruction of human embryos, the distortion and bypassing of normal reproductive processes, the cloning of human beings, and so forth are not morally permissible, no matter how useful they might be therapeutically. These questions are the substance of bioethics and will not be taken up individually here.

Within the traditional ends of medicine, the primary intention is using biotechnology to treat physical or mental disease. There is no question that curing or ameliorating a disease process will also result secondarily in enhancing a patient's quality of life. But here the enhancement lies in restoring health or relieving symptoms undermined by disease. The patient feels "better" and regains functional capacity, perhaps returned to the previous state of health or to an even better state (e.g., after surgery for a cleft palate). This kind of enhancement follows therapy and is part of its aim of therapy; it is not "beyond" therapy but a result of it.

This is different from enhancement as a primary intention. Here we start with someone who has no disease or obvious bodily malformation. The person (patient or nonpatient) is considered "normal" in the usual sense of that term yet feels dissatisfied with his or her portion in life. One feels unfulfilled, at a social disadvantage, or competitively deficient in some mental or physical bodily trait. The person may want to augment a state to what one thinks is a normal level or to something approaching perfection.

The motives, ends, and means of enhancement as a primary intention are morally variable. Some ends, such as the desire for healthy, bright, and lovable children, are understandable. If the means that bring these states about do not themselves dehumanize their subjects, they might be within the legitimate ends of medicine, particularly preventive medicine. An example would be genetic manipulation to cure or prevent a familial disease.

Conversely, many other intentions will focus elsewhere, such as on the thrills of going farther, faster, or with more endurance in athletic competition. Or enhancement might be driven by the adrenalin surge of seeing how far the human body and mind can be pushed beyond the established limits of species normality. Enhancement of this kind becomes an end in itself far beyond the healing ends of medicine in any traditional sense.

Some would extend the term *patient* to anyone unhappy, to any degree, with one's body, mind, soul, or psyche. This would "medicalize" every facet of human existence. Were physicians to accept enhancement of this kind as their domain, the social consequences would be dire. The number of physicians needed would skyrocket; access by those with disease states would be compromised; and research and development would become even more commercialized and industrialized than they have already become. Research resources would be channeled away from therapy per se. The gap in access to therapy between those able to pay for the physician's time and those who cannot would expand. To make physicians into enhancement therapists is to make therapy a happiness nostrum, not a true healing enterprise.

Yet if any significant number of physicians were to decide on ethical grounds that enhancement, as an end in itself, is not the physician's responsibility, enhancement therapy could become a field of its own "beyond" medicine. How these new therapists would relate to patients and physicians is unclear. Would they be simply those physicians willing to cooperate? Would they be persons in other fields—like sports trainers, psychologists, and naturopaths—who would attend to their own special spectrum of enhancement requests? What would these enhancement therapists do when serious, mysterious, or potentially lethal side effects appeared?

It is likely that an outright rejection of enhancement would encounter strong lay and professional resistance. The satisfaction of personal desires, freedom of choice, and a "quality life" have for many become irreducible entitlements in a democratic society. Today few want restrictions placed on their choice of enhancement. Peer pressure, the drive of a competitive society, and market pressures, inflated by "needs" generated by advertising, will convince many physicians and ethicists that it is useless to resist. Given our society's incessant search for satisfaction of all its desires in this world, many will argue that enhancement is, or even should be, part of the physician's responsibilities—no matter what the profession thinks.

The real possibilities of injurious side effects are not likely to be a deterrent. Many already see the trade-offs as worthwhile. Many already willingly sustain the expense and discomfort of liposuction, breast implants, growth hormone and steroid injections, botox infiltration, and repeated attempts at in vitro fertilization. Many eagerly pay $500 for a jar of face-rejuvenation cream. The desire to correct what unassisted nature did not provide is strong and persistent in human nature. The confluence of an ego-oriented culture sustained by social approval, peer example, and clever advertising will produce a cascade of demand.

Physicians will be drawn into enhancement practices for a variety of reasons. Some will see only good in it; some will accept it as "treatment" for the unhappiness and depression suffered by those who are not everything they want to be. Others will argue that physician involvement is necessary to assure safety and to permit better regulation of abuses. "What better way to treat the whole person?" some may add. "Isn't the patient the one who knows most about his own good?" Assertions like these suggest that failure to provide enhancement may become a breach of the physician–patient relationship or the physician's social contract.

Enhancement will also appeal to the physician's self-interest. A willing and paying clientele is certain to develop. Patients will be more eager to pay for the enhancement of the lifestyle they desire than for a treatment of a disease they did not want in the first place. Physicians can say they are doing "good" for their patients even while doing well for themselves. The possibility, and probability, of a serious conflict of interest on the part of the physician cannot be ignored. The financial motive can easily induce the physician to provide an enhancement of dubious merit or marginal efficacy. More specific, for example, is the conflict involving the team physician who is expected to do his or her part to produce a winning team. Enhancements of athletic performance are in worldwide use. Their harmful side effects are well known. Who does the physician serve? The good of the patient, the success of the team that pays his salary, or his own infatuation with athletic success?

The fundamental questions about how enhancement affects our concepts of the purposes of human life and the nature of human happiness will be buried by the more immediate demand for happiness, fulfillment, and mental tranquillity.[38] The modern and postmodern emphasis will be on effective regulatory measures, better techniques, and competent practitioners—not on ethical restraint. Restraint or prohibition beyond the prevention of abuses and harmful side effects is therefore highly unlikely. Those who restrict freedom of choice will be seen as a danger to the realization of a higher quality of life for all. Any restriction will be interpreted as a violation of the physician's obligation to respect patient autonomy or even beneficence.

Many of us will take these to be specious arguments that, if accepted, would make medicine the servant of biotechnology and erode its more traditional role in treating the sick. Counterarguments will be difficult, given the powerful vectors of change in our cultural mores. Hopes for an earthly paradise are seemingly within reach for many people who no longer believe in an afterlife. For them, extracting the maximum from personal enhancement is a seductive substitute.

Reengineering the Human Race

Creating a new or more perfect human species is biotechnology enthusiasts' most grandiose project. Whether by genetics, cybernetic mechanisms, or nanotechnology, the hope is to purge the species of its historical evils and to bypass the forces of evolution and natural selection. Its aspirations shame the Titan Prometheus into insignificance—nothing so primitive as stealing the secret of fire is involved; humanity will simply control all life processes and shape them to its will.

In this reengineering of species, physicians as physicians will have subsidiary roles, perhaps as preliminary experimentalists with small samples of the species. They may be called upon to evaluate the effects of procedures or to treat the mishaps. They may well be replaced when engineering methods are standardized and applied to the whole species. Their discretionary space in applying the knowledge of biotechnology will be severely limited by the grand design, whose configuration others will determine. Once perfected, enhancement techniques are easily shifted to a new profession to make them more widely available.

Fabricating a new species does not conform with the healing purposes of medicine or its concern for the sick. The subjects to be transformed will be mostly in future generations. They will have no choice but to be improved. It will take time for the processes of change to eliminate the last traces of the defects of the original species. Physicians will themselves be subjects, as will their progeny. Any semblance of professionalism would be erased in the utilitarian dream of a new species perfected according to a new template for human nature. Once the template is set, there will be little need for the physician's discretionary function; indeed, it will be an obstacle to the utopian dream.

Ethical Reasons for Restraint

Daniel Callahan and Leon R. Kass, two of the most thoughtful commentators on the bioethics scene today, have raised serious arguments for restraint in the use of biotechnology.[39] Some of their arguments and some additional ones follow. Our questions are posed in terms of medical good and the good for humans. Can enhancement satisfy the moral requirements of medicine to act for the good of the patient and the common good? Can enhancement serve the end of the good for humans as humans?

One ethical question currently being discussed is distributive and social justice. If enhancement is a good for humans, should it not be available and accessible to as many as possible? Costs will undoubtedly be significant, and many will be excluded for lack of means. Should physicians provide the putative benefits of enhancement for the few, who then presumably can compete more successfully with the "unenhanced?" The concept of "haves" and "have-nots," then, will extend to personal characteristics endowing one with social advantages. The current evolution into a two-class society will accelerate. Those not enjoying enhanced minds or bodies will be doomed to a life of restricted opportunity or virtual, if not actual, submission to the will of the "enhanced."

Can society or physicians condone the expenditure of limited resources for the enhancement of the few when so many in the world are impoverished, hungry, and sick? Is there not an obligation to our fellow humans to provide housing, sanitation, and pure water to the many in place of the enhancement of the few? The global implications of this maldistribution of humanity's resources must eventually be confronted.

Can a world increasingly united by communication, information, and economics peacefully tolerate such a disproportionate allocation of its resources? In this confrontation, refusing complicity in such an unbalanced resource distribution is the physician's moral responsibility.

Medical knowledge and the specialized knowledge of enhancement particularly are not the physician's proprietary possession. Physicians have the obligation to use society's resources wisely as good stewards, not as shopkeepers whose only aim is customer satisfaction.[40] Surely sanitation, immunization, and access to necessary medical care ought to take precedence over a super athlete's urges for a few more seconds off the four-minute mile or a new record for home runs in a year. The billions now being spent worldwide on enhancing athletic achievement are a shameful commentary on humanity's priorities.

A powerful argument relevant to the cloning debate is equally applicable to the enhancement question. That is the repugnance, the intuitive feeling of disgust and revulsion, many experience over the overweening hubris and self-worship that enhancement fosters.[41] On grounds of both secular and religious humanism, the gross obsession with one's self (body and psyche) that motivates primary enhancement would be unseemly at best and pathological narcissism at worst.

The protagonists for enhancement lean heavily on the priority of autonomy to rebut arguments like this. But do enhancements of memory, or intelligence, by drugs or implanted electrodes, enhance or weaken autonomy? Do they not put humans under greater external control? Predetermining our habits, behavior, or capabilities to a specific end must limit other interests, pursuits, and lifestyles. If our progeny have been genetically engineered for certain roles (athletic, intellectual, or artistic), they are excluded from other roles requiring different attributes. Is that what they would choose? Is the "willing cyborg" really willing?[42]

Does not this type of "enhancement" trade freedom of choice for submission to the hegemony of the technobiologists? As C. S. Lewis, with characteristic pithiness, puts it, "Each new power won by man is a power over man as well."[43] It would be ironic indeed if our intractable and insatiable demands for personal freedom were to end in the limitation of some of our most essential freedoms.

Then there is the question of commutative justice. Whom do we reward when a cyborg, an enhanced athlete or scholar, or a courageous soldier performs outstandingly? Is it the drug, the chip, or the apparatus? Is it the designer of the enhancing modality? The inventor of the idea? The restructured human? Who is entitled to claim credit? Who gets the prize, the citation, the medal, or the honorary degree? Will the Nobel or Olympic medallions need to be cast in multiples of six or twelve to do justice to all?

More seriously, what happens to justifiable pride in accomplishment, to merit, to virtue? It is very doubtful that primary enhancements will be good for humans. They will not make for better human beings, nor will they make us more courageous, compassionate, just, kind, or altruistic. They will more likely dehumanize us. There is, after all, no causal correlation between intelligence, athleticism, or artistry and humaneness. Indeed, enhancing some humans' preferred capacities will only enhance their efficiency in doing evil.

Will society be a better society when more of its members are taller, more muscular, more intelligent? The massive growth of sets of characteristics popular at the moment will lead to social divisiveness, devaluation, and dissent on the part of those who choose different enhancements or none at all. A society that is homogenous in sharing some vision of self-indulgence stifles progress and originality. A society with groups seeking enhancement but differing from each other in the qualities they prefer invites destructive competition.

What about moral and legal responsibility? If we find it difficult to praise artificially induced achievements, whom do we blame for failures? Is the cyborg or the implanted chip responsible for a crime? What is the moral accountability of the physician or technician who enhances intelligence that is directed at power and tyranny rather than social good? Hans Jonas and Jacques Ellul, two of the most perceptive commentators on the powers of technology and its impact on future human affairs, have raised the fundamental questions of moral accountability in particularly prescient ways.[44]

In any case, a society filled with disease-free people all living indefinitely in different enhanced states will face insuperable problems. Competition for space, resources, and opportunities for self-expression will defeat the very purpose for which enhancement was initiated. Access to resources will need to be regulated. Regulation will mean restrictions on the drive to self-determination that posthumanists so earnestly seek. The age-old questions of who decides whom, and what, to regulate will surely surface once again, and with it another dimension of the accountability issue.

The vision of a better or improved human species assumes a capacity to plan and design well enough to know which characteristics to enhance, which to eliminate, and which to never pursue. The presumption is often that those best qualified to make such choices are those who understand, and control, the biotechnological means. But there is no evidence that scientists are any better than others in deciding what is good for humans as humans. To suppose otherwise is to fall victim to scientism and its ideological pretensions. These pretensions were evident as long ago as 1963, when the earliest possibilities of genetic enhancement were beginning to be grasped.[45]

Jonas sensed the irrefutable fact that engineering a better future generation "demands an exponentially higher degree of science than is already present in the technology from which it is to be extrapolated."[46] And a lack of this requisite "higher degree of science" could lead to a massive experiment in species transformation, ending in species disaster rather than enhancement.

Jonas's prescience catches our attention when we look at two current enhancement techniques: nanotechnology and systems biology. Nanotechnology has two aims—one relatively benign, the other potentially disastrous. In the first case, nanotechnologists fabricate minuscule robotic motors that, for example, could deliver chemotherapeutic agents selectively to tumor cells. Nanotechnologists can fabricate new shapes for molecules, such as "buckyball" carbonmolecules, or carbon nanotubes with extraordinary new properties and strength. In this form, nanotechnology is a highly sophisticated expansion of materials science with many possibly beneficial applications.

The other direction for nanotechnology is more radical. It aims to make biological machines from selected bits of the DNA molecules that could be used "to manufacture just about anything man wished."[47] This form is very attractive to those who seek to totally remake human nature and give humans their dream fulfillment—that is, an ageless, disease-free immortality. It includes self-replicating molecular machines.[48]

Another version of this nanotechnology is synthetic biology, using bits of DNA to reengineer genetic circuits and thus produce made-to-order molecules. These molecules could clean up pollutants and detect explosives, or as "biobricks" could make microbes into programmable computers.[49] The problem is that these made-to-order microbes could get into the environment, multiply, and vastly alter the earth's ecology.

In any case, these and future efforts to change human nature—that is to say, the transhumanist, posthumanist projects—are well outside the ends of medicine. The moral end of clinical medicine is the good of the individual patient. Nothing in our experience thus far justifies the pursuit of such utopian aspirations. Nor is there any evidence that, were it attainable, the posthuman state would make us happier, better, or more peaceful. As Jürgen Habermas recently pointed out, there are inherent difficulties in wedding current interpretations of good science and liberal philosophy, as is attempted in the dominant notions of most biotechnologists.[50] Together, these beliefs can tyrannize as much as they liberate. Despite these warnings, a whole new field of nanomedicine is being envisioned to exploit nanotechnology for medical purposes.[51]

Biotechnology and the Physician–Patient Relationship

Whether or not medicine should embrace biotechnology and to what degree must take into account changes in the nature of the physician–patient relationship over the past half century as medicine has been altered by scientific, economic, and social pressures to an extent yet to be fully appreciated. Physicians have been urged to become clinical scientists, entrepreneurs, resource managers, employees, and advocates of society, not of individual patients. As a result, patients have been made into objects of technical interest, customers, consumers, and a drain on public funds. Although still viable with most physicians, their traditional role as healer has been seriously eroded.

As this debate continues over what it is to be a physician, we must be careful that biotechnological opportunities for enhancement do not further corrode the profession. At a time when professionalism is widely thought to be in crisis, it is imprudent to accept a new role as enhancers of personal and idiosyncratic desires. The augmentation and enhancement of every facet of what it means to be human do not guarantee that we will be more human. Being best at something, or more perfect in some trait, does not guarantee happiness or health.

Despite all this, we must recognize that prominent bioethicists of a more techno-utopian persuasion see few ethical problems—even with the most provocative techniques of brain manipulation. For example, one prominent ethicist, Arthur L. Caplan, says "I see nothing wrong with trying to enhance and optimize our brains."[52] He rejects the notion that enhancing some will disadvantage others, that brain manipulation will make us less human, or that we will all be subject to coercion to follow the lure of enhancement. Smart pills, improving memory, and downloading the content of our brains on a computer (primary enhancement) are not analogous to using eyeglasses, artificial hips, or insulin.

Questions of this type have already spawned a new subbranch of bioethics called *neuroethics* to deal specifically with the questions that emerge from the cybernetic, electromagnetic, and pharmacological manipulations of the human brain. The ethical issues of brain enhancement have special moral significance because of the close association of brain function with thinking, personhood, free will, and behavior.[53] Sophisticated methods of brain scanning seem to suggest a biological basis for mind, soul, and psyche, recalling Descartes' hope of finding the soul in the pineal gland. Though they might admit that the soul is not in the pineal gland, many take the new evidence from neurophysiology to support a monistic materialist reduction of human nature to an epiphenomenon of the complexity of its molecular arrangements.

This is the modernist and secular humanist answer to the Psalmist's question, "What is man?" (Pss. 8 and 144). But is it a defensible response to the request of the President's Council on Bioethics for a "richer bioethics" grounded in a better understanding of what it is to be human?[54] For the secular humanist, the materialist monist reduction would suffice as the final reference point for judgment and inquiry, not just for enhancement but for all bioethical issues.

This idea of humanity is of course directly in conflict with the philosophical and theological anthropology of Christianity. The secular humanist and Christian humanist anthropologies are thus set in more intense opposition than ever before. This dissonance is further aggravated by recent attacks on a central moral principle of Christian ethics, namely, the principle of human dignity. One bioethicist, Ruth Macklin, has summarily affirmed that dignity is a "useless concept." She argues that it is "hopelessly vague" and too closely related to its religious origins, particularly the Roman Catholic. Another author, Matti Hayry, argues that there are four competing and conflicting meanings for "dignity"—namely, the Christian and the Roman Catholic, the Kantian, the genetic, and that expressed in the UN Declaration of Human Rights.[55] This author argues that should one or the other of these definitions become dominant, the term *dignity* would become antiegalitarian.[56] Both these ethicists' opinions are extrapolations of a strong trend in bioethics to discredit any notion of a unique moral status of human beings except for autonomy. And the capacity for self-governance is proposed by both these ethicists to encompass what they consider to be a more important basis for bioethics than dignity.

This is not the place to expose the speciousness, logical contradictions, and moral poverty of such distortions of so fundamental a concept as human dignity (chapter 3 of this volume describes the criteria for assessing the foundations of a concept of dignity). These assaults on human dignity, if successful, are sure to frustrate any notion of a "richer" bioethics. In its place, we can expect a morally conflated, utilitarian, hedonistic cult of self and selfishness.

Enhancement and the Christian Physician

Up to this point, consideration of the ethical issues related to enhancement has drawn on what is ascertainable by human reason, and in the context of today's bioethics, which admits of no agreed notion of what it means to be human. For Christian physicians, the discussion starts elsewhere, precisely where the secular liberal view breaks off—namely, with an idea of humanity, a philosophical and theological anthropology upon which all morality is based.

We have seen an open confrontation with Christian anthropology, which offers the only real hope of moral restraint on the rising tide of infatuation with individual and species enhancement. What is fundamentally different about the Christian notion of humanity is its firm grounding in a transcendent order of Creation and morality. The source of humanity's dignity and its special moral status and accountability is God, who reveals Himself through the scriptures, church teachings, tradition, and practices of Christians. These are reflections of Divine Law, only part of which is open to human reason unaided by revelation.

Thus, human reason can provide the basis for a philosophical anthropology that will always be incomplete without the insights of a faith commitment. Nothing in the history of human thought supports any notion that unaided reason can fathom the full mystery of humanity. Humanity's spiritual nature and destiny can be grasped sufficiently, however, to provide a reliable guide by which the moral quality of enhancements and their relations to the good for humans can be judged. The more we probe, the deeper the mystery becomes, as the queries of Ecclesiastes and Job make clear. But also the more we probe, the more we appreciate that the human soul is more than the result of the emergent complexity of matter.

Men and women of faith will know that the taint of Original Sin cannot be eliminated by microchips, nanotechnology, or cyborgian metamorphoses. Neither will they seek salvation in utopian redemption fashioned by humanity's hubris. Rather, they will know that the gifts of science and technology are good only if they make humanity good and more cogniscent of God's creative presence in each wonder science uncovers. Believers are not likely to surrender their faith or fate to biotechnologists or to sell them for a mess of enhancement pottage.

Medicine and the Four Levels of Good

For the Christian, enhancement need be neither feared nor adulated. The possibilities of the new biology are to be used for human good but not for humanity's deification. Medicine as medicine can apprehend only the medical good, that which is directed to the relief of disease and the suffering caused by disease and susceptible to application of medical knowledge. That is the only good to which medicine can address itself authoritatively. There are three other levels of human good beyond medicine. In ascending order they are (1) the personal good perceived by the patient within the existential context of one's own life, (2) the good of the patient as a human being, as a being of a certain natural kind, and (3) the spiritual good—that defined by the spiritual destiny of a person, which is union with God.[57]

To heal, medicine will need to take account of all four levels, though its particular domain is in the lowest good—the medical good. Medicine must rely on the psychosocial sciences and the humanities to comprehend the personal good; on philosophy to comprehend the good for humans as beings of a natural kind; and on religion or one of its substitutes for that hunger for immortality, which the quest for perfection in this world so frequently obscures.

Resisting the Spiritual Temptation of Super Enhancement

For the Christian, primary enhancement will always be a temptation to pride, the same temptation to which Lucifer and Prometheus succumbed. Redemption of humanity cannot come from humans but only from God. Faith cannot rest in the promise of utopia in this world, but neither will it despair. With faith come hope and charity, and happiness in the love of God and humanity in even an imperfect world.

Super enhancement and the new "posthumanism" are disastrous spiritual distractions from the reasons for which we were created. These are challenges to God's sovereignty over us, to His reasons for our creation and a denial of the indispensability of the Atonement and the Resurrection. The perfection we are enjoined to seek is the perfection of our spiritual and moral lives—a perfection we can pursue only asymptotically here on Earth. Yes, we are stewards of our bodies. Yes, we should use technology wisely and well. In this we follow Jesus' own healing example and mission. But healing was always within His larger mission. It was never to pursue the pagan cult of idolization of the body, which so dominates today's culture.

The Christian is neither a Manichaean ascetic nor a Dionysian narcissist. Jesus was incarnate in a body like ours. Our guide to the proper use of our bodies lies in a Christological theology of the body.[58] That theology is increasingly at odds with the atheistic perception of a perfectible human existence without God. As we enter the transhumanist and posthumanist era, the possibility of some synthesis between the scientific and Christian view of humanity seems remote.[59]

The Christian view is clearly at odds with biotechnological optimism. Human knowledge is a good because it is at heart a search for God. But it can be used as well to destroy humanity. Martin Rees, the royal astronomer of the United Kingdom, suggests that the odds are no better than fifty–fifty that humanity will survive this century, largely because of the possibility of mass casualties resulting from the misuse or abuse of our new powers.[60]

Humanity is more than the result of natural selection and genetic endowment. We are destined for more than satisfaction of our personally defined hedonistic needs. Paradise for Christians is not a human-made, earthly paradise fashioned by chemical or biological prestidigitation. It is union with God and the joy of fulfilling His will, not ours. Only in that final union with God, in a body glorified by the resurrection, can the insatiable hunger that drives humans to enhancement be satisfied.

Ellul recognized the inevitability of a "head on collision between man who wants to be himself and God who also wants man to be himself." As he goes on to explain, "the difficulty is that 'himself' does not mean the same things in both cases; in fact, the one meaning contradicts the other."[61] This is also precisely what the more outspoken proponents of scientism have also reasserted.[62] This is a powerful negation of the reconciliation so many have hoped for.[63]

Those who promote or demand enhancement suffer from what Robert Burton, almost half a millennium ago, called "melancholia," defined as "a state of mind in which man is so out-of-touch with his environment that life has lost its sweetness."[64] Burton's antidote was "an active, unselfish life of devotion to others." Melancholia is not a disease to be cured by drugs or implanted computer chips. It is an illness of soul for which there is only one remedy: the redemption of humanity by God-made-man, not the redemption of humans by humans.

Humanity's inborn thirst for perfect happiness is the thirst for God. Either it will be satisfied by the love of God in this world and union with Him in the next, or it will be filled with the substitutes of endless attempts to better the Creator.

Conclusion

Christian physicians should neither fear nor adulate biotechnology. Like Sirach in the Apocrypha, they should realize:

> From God the doctor has his wisdom. . . . He endows men with knowledge to give glory to His mighty works through which the doctor eases pain and the druggist prepares his medicine. Thus God's creative spirit continues without cease. (38:2, 6–8)[65]

It is too early to attempt to delineate specific boundaries for the use of biotechnology, and especially enhancements, by Christian physicians. Clearly, enhancements of the lives of patients as a result of the treating of disease are well within the ends of medicine. The only provision is that the means used do not violate moral constraint. Indeed, in this category, there is a moral obligation to bring to bear whatever knowledge the physician possesses that will cure, heal, mitigate, or prevent disease.

When enhancement is the sole intention of the use of biotechnology, when there is no disease present but only the desire to pursue perfection, immortality, super performance, a competitive edge, and so forth, there seems little justification for physician participation and good reasons for morally excluding it. When the motives are patently pride, narcissism, or pursuit of "happiness" or immortality, Christian physicians cannot justify cooperation. Where the dangers of side effects are also present, participation would be precluded by the end of medicine, which is the good of the patient—even when the patient is willing to run the risks of harm. In any case, Christian physicians will examine these questions from the perspective of the Christian idea and image of humanity—that foundation toward which others may tend but from which the Christian begins.[66]

Pope John Paul II put the whole question of Christian belief and technology in strongly personal terms:

> It is essential that we be convinced of the ethical over the technical, of the primacy of the person over things, of the superiority of the spirit over matter. The cause of the human persons will only be served if knowledge is joined to conscience. Men and women of science will truly aid humanity only if they preserve "the sense of transcendence of the human person over the world and of God over the human person."[67]

CHAPTER 7

Conclusion: Toward a Foundation for Biotechnology

It's knowing what to do with things that counts.

—Robert Frost, "At Woodward's Gardens"

IN the preceding chapters, we have presented the background considerations necessary to inform the theoretical foundations of biotechnology. Our goal in this conclusion is to develop these concepts into a series of principles, expressed through questions, that we hope will help assess biotechnologies philosophically, theologically, and practically. We do not in any way claim that our thoughts are a complete systematic philosophy or theology of biotechnology. Although a comprehensive treatment of these issues will have to be reserved for another volume or series of volumes, we hope that this work will begin a discussion that must take place for all of us to think critically and respond responsibly about the challenges raised by both existing and future biotechnologies.

Questioning Technology

There are many questions that we must ask both collectively and individually of any technology, not least of biotechnologies. These questions should probe not only what the technology does but how it is developed and produced. The following list is representative of the kinds of questions that must be answered before we embrace new technologies.

Question: Does the technology assist us in fulfilling our stewardship responsibilities?

- Does it treat natural resources responsibly within the constraints of good stewardship of our common resources?
- Does it create ecological problems or peril? For example, does the technology produce land, air, or water pollution or ecosystem imbalance?

As we explained in chapter 2, the Christian worldview understands humankind's place in the order of Creation as one of responsible stewardship. This stewardship necessarily entails consideration of ecological concerns and does not give license for the irresponsible destruction of ecosystems or the poisoning of our commonly shared environment. Human flourishing is dependent on environmental considerations. A well-worn canard is that the Judeo-Christian tradition is the root of the ecological destruction attendant with modern industrialization. Pannenberg has addressed the discordance between this myth and the real worldview underlying ecological irresponsibility:

> The criticism of biblical anthropology that blames the giving of dominion in Genesis 1:28 for the unrestricted exploitation of nature by modern technology and industrial society, and for the resultant ecological crisis, must be rejected as without merit. Modern industrial society has its basis in modern secular culture, which . . . cut off its historical roots in Christianity. Emancipation from religious ties and considerations . . . had been one of the presuppositions of the autonomous development of economic life in the modern age. . . . But this world is still the Creator's, and God's will as Creator is still standard for the dominion we exercise as God's image.
>
> This dominion, then, excludes arbitrary control or exploitation. . . . Because the world of nature is still God's, . . . our self-glorious misuse of the power we have been given by God rebounds upon ourselves and plunges us into ruin. In this sense we may view the ecological crisis at the end of the modern age of emancipation as a reminder that God is still the Lord of creation and that human arbitrariness in dealing with it is not without limits or consequences.[1]

Environmental impact must therefore be a necessary consideration in the evaluation of any technology.

Question: Does the technology facilitate healing or restoration from disease or disability, or is it for reengineering (so-called enhancement)?

- Does it foster or inhibit community development?
- Does it promote a community that values and accepts all individuals regardless of their attributes?

As a possible negative example, a Carnegie Mellon University study of 169 individuals over one to two years of time on the Internet found an overall negative trend in authentic human interaction.[2] More time spent online resulted in decreased communication with immediate family members, a shrinking of the subject's social circle, and increases in self-reports of both loneliness and depression.

Question: Does the technology require or promote the commodification or destruction of human life? Does the technology demean, debase, or degrade individuals?

- Does the technology require or reinforce diminished views of human life, human value, and the human being?

Some philosophers, theologians, and scientists have gone so far as to reject the given status of human beings, declaring *Homo sapiens* to be a transitory species on the way to being posthuman. They prefer a technologically reengineered being of some kind.[3] For instance, in *Technology and Human Becoming*, Philip Hefner declared:

> God is a participant in the technological process, since the purposes of God are now embodied through technology and techno-nature. . . . If the techno-human, the cyborg, is created in the image of God, what does that tell us about God? Technology is itself a sacred space. . . . Technology is about the freedom of imagination that constitutes our self-transcendence. . . . Technology is the shape of religion, the shape of the cyborg's engagement with God.[4]

Similarly, in *Beyond Cloning: Religion and the Remaking of Humanity*, Ronald Cole-Turner asks:

> Will nature transcend itself again via technology and give . . . God a transhuman species more intelligent, more spiritual, more loving, more creative, poetic, musical, more adept at praise, more generous, more able to glorify and enjoy God? . . . I have to . . . seriously entertain the possibility that our technology can and should be seen as instruments in the creator's hands, by which God will continue to sculpt Adam's clay, refashioning life in the universe so that it might give greater glory to its master.[5]

Question: Does the technology primarily appeal to our basest inclinations?

- Does it appeal to our pride by fostering materialistic, intellectual, and/or physical competition?
- Does it appeal to lust and the sexual commodification of human beings?
- Does it promote the enslavement of individuals to the fickle whims of fashions, particularly in terms of body modification?
- Is the technology a tool for the expression of anger and hatred?

Question: Is the technology a vehicle to promote our own narcissistic self-absorption?

Certain technologies to address infertility, for example, may correct certain anatomic and physiological deficiencies and may be considered legitimate interventions. Yet beyond those, extreme measures are being pursued that create excess embryos or place the woman's health at risk solely to produce a genetically related offspring, something difficult to justify in any legitimate moral calculus. Parenting is not primarily about the parents. It is about the children. There are many millions of orphans around the world, desperately longing for parents to love them. Yet assisted reproductive technologies (ARTs) more often appeal to our narcissistic beliefs that our genes are better than anyone else's, or they promote the fiction that the child who is not genetically related cannot be loved at the same depth as a biological child. Parenting is a vocation, a calling, requiring selfless commitment to the good of the child. ARTs, conversely, have a tendency to commodify the child, turning him or her into a project, a goal, or property to add to the list of achievements or acquisitions of one's life.

Question: Does the pursuit or use of the technology make just use of resources?

For example, would simpler, less costly technologies provide a more substantial benefit overall, such as reducing infant mortality, treating devastating diseases affecting much of the planet (e.g., malaria), or providing sanitary conditions such as potable water? It seems difficult to justify, for example, the $7 billion spent in the United States in 2002 on noncorrective cosmetic surgery (e.g., 350,000 nontherapeutic breast augmentations) when faced with the enormity of the infant suffering and death still remaining in the United States, let alone around the world.[6]

Increasingly, adults are seeking cosmetic surgery, primarily to maintain competitiveness in the marketplace. In a survey of cosmetic surgery patients, 33 percent of the male patients and 19 percent of the female patients pursued body modification for work-related reasons.[7] This is a prime example of *technicism* in our society, and it represents a widespread social pathology.

The images of fashionable "beauty" are carefully crafted by industry in such a way that only a scarce few do not require, according to the technocrats, significant alterations in hair color, skin tone, body, or feature shape. Many forget that the beauty myth is created by people whose job it is to sell products and services. For example, in the 1980s one cosmetic company ran a series of television and print ads using an attractive actress who cooed: "Don't hate me because I'm beautiful."[8] The obvious message was, of course, "You should make yourself beautiful, too, by using our products." However, the real message of the campaign was, "You should hate yourself for being less attractive than our airbrushed model." Marketing companies know that by inducing this self-loathing, this envy of another, they are persuading the consumer to consider buying the technology of aesthetic transformation.

What is increasingly problematic is that the images portrayed can no longer be achieved simply by the use of cosmetics. To attain the level of beauty now proffered will require a physical restructuring of the body itself. An *Esquire* magazine article once ran a piece on the actress Michelle Pfeiffer titled "What Michelle Pfeiffer Needs . . . Is Absolutely

Nothing."[9] The implication was that this beautiful actress was the perfect physical model, the standard of aesthetic perfection. However, the pictorial image of this paragon of physical superiority was significantly modified to produce the purported perfection, including complexion clean-up; softening of the eye, smile, and ear lobe lines; addition of color to the lips and hair; trimming the chin; removing neck lines; and adding hair on the top of the head. At least there was a real person as a base in this instance.

The future, however, will be even less generous to mere *Homo sapiens.* One can now find the Miss Digital World contest on the Internet. The site effuses:

> Every age has its ideal of beauty, and every age produces its visual incarnation of that ideal: the Venus de Milo, the Mona Lisa, the "divine" Greta Garbo, Marilyn Monroe. . . . Miss Digital World is the search for a contemporary ideal of beauty, represented through virtual reality.[10]

As one examines the contestants, it becomes clear that few, if any, members of *Homo sapiens* will be able to attain the morphology of these avatars without surgical modification. Is this really how we should view ourselves and one another? Should technologically enhanced humanity truly become our aspiration? Are techno-aestheticians the next stage in vocational evolution?

Our society also devalues the older among us, preferring frenetic youthfulness to wisdom, experience, and stability. This is folly. Yet in our technicistic mentality we "solve" the problem, not by repudiating the arbitrary messages of industry and instead building a community where all are valued and accepted but by using technology to mold us into someone else's image of what he or she wants us to look like today. And the statistics demonstrate that we follow along in droves, failing to remember that the current norms are going to change arbitrarily and that we will again be found unsatisfactory by the "aesthetic" elites. Perhaps this is why Jesus instructed his disciples not to worry about what they would wear (Matt. 6:25; Luke 12:22).

In 2004 the American Society for Aesthetic Plastic Surgery issued general suggestions for "improving quality of life" throughout the life span.[11] These included beginning Botox injections, with a tummy tuck and breast lift, after childbirth in the thirties; lipoplasty for double chin and fat deposits, along with injections for spider veins, in the forties; a facelift, eyelid surgery, rhinoplasty for drooping nasal tip, and lip augmentation in the fifties; and more Botox, skin resurfacing, chemical peel, fat injections, and repeat facelift and upper-arm lift in the sixties. One can immediately suspect that the only lives really "enhanced" by this program will be the cosmetic surgeons making phenomenal profits from all this unneeded whittling. The reader should also remember than none of these interventions is without risk, and some come with severe, disfiguring complications.

We should frequently recall the observation made by the historian and philosopher of technology Lewis Mumford in *The Conduct of Life* in 1951:

> Man's constant re-shaping of himself, his community, his environment, does not lead to any state of equilibrium. Even the process of self-perfection implies the projection of a self beyond that which may be momentarily achieved. . . . And tempting though it may be to do so, one must not confuse the good with what is socially acceptable, or that which promotes the adjustment of the self to the group or the community. Pragmatists and totalitarians have both made this radical error; by their insistence upon conformity to an external pattern, whether imposed by authority or by a mechanistic apparatus, they have proved hostile to creative processes that have a subjective and internal origin.[12]

Question: Does the technology promote genuine human flourishing or does it more likely promote technological and economic imperatives? Must we adapt to the technology, or was the technology designed to adapt to human nature and human needs?

The motto of the 1933 Chicago Century of Progress World's Fair was "Science Finds, Industry Applies, Man Conforms." The past century illustrates well the consequences of placing ideology, science, and industry above human needs. The eugenics movement, which took Anglo-American society by storm intellectually, theologically, and socially, is a clear example of the tragic consequences of techno-utopian thinking.[13] The eugenics movement also illustrates the limitations of naive science in making claims about humankind's nature and destiny. The products of the scientific method will always be incomplete in and of themselves for explaining our nature, because the process of science deliberately excludes consideration of the most important aspect of our being—that we are created beings.

Science is but a tool and only one means of discovering truth. Consequently, scientific theories alone are inadequate as guides to structure society and our vision for humankind. Technologies should always remain the tools of human beings, not vice versa. Schuurman cautions:

> When Christians accept technology only positively and uncritically—and, for instance, reduce the cultural mandate to a technological mandate—they are blind to what is really going on. . . . In reality they have closed their eyes to the central fact that human autonomy is concentrated in the scientific-technological control of everything. . . . People-without-God are trying to master God's creation. They want to realize a worldly paradise, a technological paradise. In the meantime the perspective of eternity is lost. Heaven is closed.[14]

Question: How much additional technology is necessary to produce, maintain, or safely constrain/contain the technology?

These concerns are of significant importance when we consider environmental impact. Although the grey-goo scenario of a widespread ecologic catastrophe caused by destructive nanotechnology is now downplayed as an unlikely concern by some, even by Eric Drexler (who in 1986 first raised the possibility of rampant nanotechnology-based ecophagia in his *Engines of Creation: The Coming Era of Nanotechnology*), large-scale hazards are still possible.[15] Engineered bacteria and viruses are already a reality. Engineering organisms or mechanisms to create widespread destruction, toxin production, and other malicious activities are not in the least farfetched. The unintended effects of genetically engineered plants already pose a significant ecological problem. Consequently, we must prospectively discern what resources will be necessary to protect against such hazards as our technology makes these things more and more feasible. We should ask:

- How will nanoscale devices be detected and inactivated as need be?

- Should individuals and corporations wishing to release such devices into the environment, be it in vitro or ex vivo, be required to demonstrate proof of containment, detection, and inactivation to receive a license for testing and/or receiving a patent?

These questions are of the type that our communities must raise. We must engage these technologies prospectively because the potential for harm is too great to respond only reactively. To ask these questions is not antiscience, anti-industry, or antiprogress, as techno-utopians often rail—it is common sense and good stewardship.

Thus far, we have established what we believe are the necessary presuppositions for a meaningful engagement with biotechnology. We have begun to ask appropriate questions of technology. We now delineate the pillars of our own foundation for biotechnology.

Presuppositions for Engagement

We must begin with the affirmation of a creator of everything. A theistic worldview is consistent with the scientific, historical, and subjective evidence available,[16] and technology is an activity in which humans exercise freedom and responsibility in response to the creator God (cf. chapter 5). God's existence affects the human relationship to the natural world because humans are responsible as stewards of God's creation. As stewards, we are not free to do just anything we please to one another, ourself, or the environment. As accountable beings, we are to embrace both ourselves and our environment nurturingly.

We also affirm that the biblical account is the best guide to understanding the nature, problems, and ends of human life. The mechanism God used to create human beings—whether by special creation or evolutionary mechanisms—is irrelevant in light of the reality that God has created human beings for relationship with God and others. Human beings are bearers of God's image by God's choice and appointment (cf. chapter 3). Imaging God is a gift, an office given to us by God, not a set of functional attributes that human beings may define, or define away, according to our fickle and selfish ends. God chose to become incarnate as a human being in His Son, Jesus of Nazareth, further distinguishing human beings from the rest of creation as worthy of special consideration among all other forms of life. Furthermore, for Christian theists the belief in the historical reality of the Resurrection of Jesus Christ means that God, in Christ, has sanctified humanity.

As all human beings—regardless of age or level of development, health, disability, or status—are God's imagers, each is worthy of respect and protection. This view is not idolatrous, placing human life above God. God remains sovereign and is, alone, worthy of worship. This affirmation entails, however, that no human is to be commodified, demeaned, exploited, harmed, or destroyed for the sake of the goals and desires of others.

Human beings are also distinct from human tissues. A human being is a complete biological entity capable of self-development, following a trajectory toward human adulthood and/or maintaining that state via homeostasis until death. Consequently, gametes, adult stem cells, and other tissues that are not capable of developing into an adult human being with normal biological support in utero, or post partum, are not human beings. Because this is true, biotechnologies that would destroy or injure human beings in their development or function should be resisted or banned.

Human beings were created for community and communion, with God and with one another. John Zizioulas has stated that "There is no true being without communion; nothing exists as an 'individual' conceivable in itself. . . . The person cannot exist without communion."[17] We are bound together in the human family and bear responsibility for the treatment of one another. Human decisions about technology are not autonomous. Our choices affect the utilization of resources, the nature of our communities, our treatment of our own bodies and selves, and our treatment of one another. These choices communicate and actualize our values and reveal how we value others. Our choices to create, purchase, or use certain technologies either will promote our development and respect for one another as images of God or will demean that image, separating us from the communion of human fellowship and from God.

The fundamental problem of humankind is not physical or mental inadequacy, but sin (Mark 7:21–23). Sin has driven a wedge between human beings and between each human being and God. Sin brings the pain and isolation of human life—the envy, greed, lust, hatred, pride, and selfishness that rob us of joy and fulfillment in our lives—and diminish the intimacy of communion. Some technologies may ameliorate the effects of sin, but no technology can eradicate the reality of sin. Humankind has pursued its own supposed solutions throughout the history of our species, often ending in the further degradation or destruction of human life, further alienation from one another, and an increasingly narcissistic descent into fear, anger, mistrust, and self-consuming greed. The history of the twentieth century is primarily the story of one utopian secular quest for the perfect human and the perfect society after another, and it ended as the bloodiest, most starkly barbaric century in human history. The tragic realities of communism and fascism and the excesses of liberal democracy, with its emphasis on individual autonomy and libertarianism to the exclusion of greater human flourishing, have all been replaced and/or joined by the equally flawed faiths of scientism and technicism.

Facing Scientism and Technicism

As defined by Stenmark, *scientism* includes a number of beliefs and attitudes that elevate science to a supreme level of epistemic and ontological authority and, in its most extreme manifestations, to an almost godlike status.[18] These include the belief (1) that all bodies of knowledge should or must develop the methods of the hard scientific process to achieve validity and (2) that, thus, science is the only really valuable part of human knowledge and culture. Scientism is manifested in the view that all nonacademic areas of human life can be reduced to science, the belief that the only reality that exists is that which science can access and describe, and with it the claim that only information obtained via scientific method can be rationally accepted, the belief that science alone can explain morality and should dictate ethical standards, the claim that science alone can explain and replace religion, and—not surprising, given the grandiose claims thus far—the faith that science can eventually solve all human problems. Given the very nature of scientific investigation, with the requirement for the repeatability of observations, this solitary approach to epistemology excludes all unitary, historical events, at least prior to video and audio recording methods (and those may now be questioned given the ability to modify data electronically), yet these events have just as much an impact on our being and culture as technology.

Science is a wonderful tool, and it has provided valuable insights into the nature of physical reality, allowing us to enjoy God's genius; but it is not the end all of knowledge and human experience. In fact, science makes a pretty poor god; and scientism is a weak, ultimately unsustainable belief system that cannot produce true human flourishing. No system that categorically denies such a large portion of reality and human existence can do so.

A close cousin of scientism is *technicism.* Monsma observes that technicism "reduces all things to the technological, seeing technology as the solution to all human problems and needs." He continues:

> Technology is a savior, the means to make progress and gain mastery over modern, secularized cultural desires. This "technology for technology's sake" is the ultimate form of technicism. More specifically, technicism is marked by three key characteristics or beliefs: (1) technological change—the development of ever more complex, ever more sophisticated technological objects—is inevitable; (2) such change represents progress, leads to improved conditions for humankind; and (3) there are technological solutions to the problems engendered by technological change.[19]

The Dutch philosopher Egbert Schuurman adds:

> Technicism entails the pretension of human autonomy to control the whole of reality: man as master seeks victory over the future; he is to have everything his way; he is to solve all the problems, including the new problems caused by technicism; and to guarantee . . . material progress. . . . One can argue that . . . the main trend of Western philosophical thought is best characterized as *thinking through technology.* This means that science and rationality in general are distorted because they have been used as technical instruments in the service of technological power.[20]

Misplaced Faith in Human Artifice: Some Consequences

The lessons of human history predict that misguided faith in human artifice will end in tyranny and the denigration and commodification of human life. Dietrich Bonhoeffer understood this point clearly when he observed in his *Ethics*:

> Defiant striving for earthly eternities goes together with a careless playing with life, anxious affirmation of life with an indifferent contempt for life. Nothing betrays the idolization of death more clearly than when an era claims to build for eternity, and yet life in that era is worth nothing, when big words are spoken about a new humanity, a new world, a new society that will be created, and all this newness consists only in the annihilation of existing life. . . . Life that makes itself absolute, that makes itself its own goal, destroys itself. Vitalism ends inevitably in nihilism. . . . [The] absolutizing of life as an end in itself . . . destroys life. . . . We call this error the mechanization of life. Here the individual is understood only in terms of its use to an all-controlling institution, organization, or idea.[21]

Though technologies do not possess salvific potential, and may be used as a means for evil, they nevertheless may be a significant force for good. An example of the beneficial use of technology has been medical technology. Yet even those technologies may be misused. Healing, the restoration of form and function to species-level norms, is an activity clearly demonstrated by the life of Christ, and it is one that we should emulate. However, it is quite another thing to attempt to reengineer human beings and manipulate the human species into other than what we were created to be.

Reengineering or Enhancing the Human Order

As explained in chapter 6, reengineering, or so-called "enhancement," is a repudiation of God's means to achieve a truly human future. All human beings lament the realities of pain, suffering, and mortality. Christian believers know that the solution to these realities does not ultimately rest in a technological "fix." No amount of evolutionary tinkering, or self-engineering, will achieve physical immortality. Rather,

Christians understand that the bodily resurrection has been promised as the door to physical immortality. The bodily Resurrection of Jesus along with his Ascension to God's right hand have been given to the world as verification of this hope.[22]

Unavoidable Liabilities and Limits

Transhumanist technologies, however—whether they involve cybernetics, nanotechnology, mind uploading, or genetic engineering—all remain enslaved to physical corruption. All these technologies will have their liabilities (e.g., the continual need for a reliable power source, software viruses, and mechanical decay). And we may predict that the application of these technologies will result in an upgrade race, or even a war, to maintain a physical or competitive edge. Just as we have to replace our computers every few years, so human-produced advancements will make our reengineered selves frequently obsolete. Neural interfaces seem particularly troublesome; and while the idea of brain chips may, on one level, sound appealing, we should remember that as we change the hardware and software substrate of our thought processes, our thinking will necessarily change as well.

Furthermore, as we all know by experience, one of the major problems with the high-technology world today is incompatibility. One company's operating system is unable to work seamlessly with the same product of a different iteration. This problem could be devastating to humans with implantable devices. So while one can justify the risks of neural implants in the hopes of restoring lost function, is it really wise to expose one's mind and body to these risks for some elusive, never satisfied goal of "enhancement"? The wisdom of scripture declares otherwise: Whoever would save his life will lose it (Matt. 16:21–26; Mark 9:31–36; Luke 9:21–25; John 12:23–26).

Constructive Human Enhancement and God's Purposes

The Christian's goal and destiny are to become like Jesus. Any constructive augmentation of any human being is to be undertaken by God and for God's purposes, not ours. In 2 Corinthians 3:18, we read: "And we, who with unveiled faces all reflect the Lord's glory, are being transformed into his likeness."

Scriptural Guidance

Contrary to our culture's narcissistic orientation, life is not just about "me." In 1 Corinthians 4:7, Paul states: "We have this treasure in jars of clay to show that this all surpassing power is from God and not from us." In the letter to the Galatians (2:20), he further declares,

> I myself no longer live, but Christ lives in me. So I live my life in this earthly body by trusting in the Son of God, who loved me and gave himself for me.

Human beings are to be good stewards of their bodies. Romans 12:1 states: "I urge you . . . in view of God's mercy, to offer your bodies as living sacrifices, holy and pleasing to God—this is your spiritual act of worship." Paul adds, in 1 Corinthians 6:19, "Do you not know that your body is a temple of the Holy Spirit, who is in you, whom you have received from God? You are not your own; you were bought at a price. Therefore, honor God with your body."

Reengineering the body does not add one whit toward the goal of holiness or toward the realization of being a living sacrifice to God. In fact, given most of the reasons people desire such modifications (rejection of their natural limitations, greed, competitiveness, lust, ambition, pride), the modifications would be a desecration of God's imagers. To attempt to wrest control of our nature and destiny in this fashion is to repeat the same mistake of our forebears in the Garden of Eden. The temptation is identical: to be like God. It is just the shape of the fruit that is different. As much as we may think we need to improve on the human design, *Homo sapiens*, with all its limitations, is the deliberate handiwork of God, sanctified by the Creator of the universe through the incarnation of God in Christ.

Pursuing Christian Perfection

What are we to make, then, of the commands in scripture to pursue perfection? (Lev. 11:44–45; Deut. 18:12; Matt. 5:48). Each of these verses is an encouragement to *moral* perfection, the perfection of *love*. The two greatest commandments are (1) to love the Lord God with all our heart, mind, and strength; and (2) to love our neighbor as ourselves (Matt. 22: 37–40). There is nothing mentioned here about bodily and cognitive attributes. The issue is one of the affections and the will, aspects that no amount of scientific and/or engineering acumen or technological ingenuity can address. God is the one who has the authority to make the profound changes required in these areas.

Christ's Healing: Why Did He Not "Enhance"?

The most significant challenge to the reengineering agenda, however, is Christological. Jesus is the Christian's ultimate example, in both His nature and His behavior. Jesus healed as a significant part of His ministry while on earth. Indeed, healing has been a significant aspect of the life of the church from its inception. The development of hospitals and many traditions of compassion trace their origins to Christianity and to Christ's parable of the Good Samaritan. Christians are called to be agents of healing and restoration, and they should embrace those technologies that assist in that goal.

But a key question must be asked: Why did Christ not "enhance" or reengineer? Doubtless he could have. Why did he not augment his disciples, making them the fittest, most attractive, and most intelligent of men? Why did he not make them impervious to pain and death? Why were their lives not extended to 200 years or beyond? The scriptural silence is deafening. Someone may retort that making a normative claim from scriptural silence is an interpretive error. Yet it is not just the silence of scripture but the silence of more than two thousand years of church history that must be faced squarely—a history that has seen the continuation of the ministry of healing but has never aspired to human augmentation in the way it is being considered today.

There may be three reasons that Christ did not augment or "enhance." First, the apostles and disciples did not, as we do not, need to be altered to demonstrate and fulfill the Gospel. Numerous passages in scripture attest to the demonstrative power of Christ's love in the face of human weakness (Rom. 8:26; 1 Cor. 15:42–44; 2 Cor. 12:9, 13:4).

The message of the Gospel is that love—not performance, competition, or enhancement—is what is important to God and should be most important to us.

Second, Christ knew that our fundamental problem is sin and that given our sinful proclivities we would pervert enhancements to make them serve us, not Him. Superior abilities often lead to superior arrogance. The story of Samson (Judg. 13:24–16:30) is a clear example of such an abuse of divinely granted enhancement. Arrogance leads to self-reliance. God wants us to be His instruments working out His plan for history in His wisdom. As has been true throughout history, ability without wisdom is often the road to suffering. What the disciples needed, and what we need today, is a relationship of total dependency on God and transformation by the Holy Spirit from within.

If the reader thinks this is too simplistic, we invite an honest survey of one's own attributes as they presently exist. Are one's gifts used to their fullest potential to serve God? Or are they used for self-gratification? If one's attributes are not being used to their fullest for God's service now, what assurance does one have that reengineered abilities would be used for God's service rather than self-aggrandizement? Is better memory, a fuller bust line, increased muscular strength, sharper visual acuity, or a different this or an increased that really going to make anyone live the Gospel more fully?

Third, augmentations or enhancements would have a significantly negative effect on the building of holy community and on nurturing communion where all have a part and a contribution, no matter how small. The more we strive for self-sufficiency and bodily perfection, the more we repudiate the community God designed for us and expects of us.

After all, a teleology is at the heart of both the Jewish and Christian faiths. Eschatological hope enables one to persevere in the midst of the frailty and corruption of the world and its inhabitants. From the Christian perspective, of course, the Resurrection of Jesus from the dead and the kind of immortality achieved through his Resurrection provide the pattern for those who follow. The *telos* for humans is, therefore, not *post*human but *fully* human, without the presence, power, and penalties of sin. Human life, as it was intended to be, will be attained.[23]

Although it is often difficult to understand—bombarded every day as we are with messages that this or that will finally bring about the fulfilled life—a stronger body, better memory, an IQ of 200, and better looks will not bring the satisfaction, contentment, joy, and peace that we seek. We can find that peace only as we return to the authentic open and loving relationships for which we were created, with God and with one another. As long as we allow industry and the media to interject false hopes and fuel our sinful desires, we will be unable to achieve genuine human flourishing. Biotechnology is produced by an industry, an industry just as greedy, just as profit seeking as any other. Though many of the products created by biotechnology will be of great value in healing, and for this we are genuinely thankful, as an industry they will prey on our fears, and upon us, to achieve their profit goals. The profit motive drove the enslavement of humans of a particular race two centuries ago, and the profit motive is driving the quest to create, enslave, and destroy humans of a particularly young age today.

Next Steps: Engaging the Theological Community

The challenges presented by advancing technologies, particularly biotechnologies, are growing almost exponentially. Yet at this time, at least in the United States and much of the West, we are theologically ill equipped to address these challenges either individually or collectively. One of the major deficiencies lies in the fact that theology has too often become an arcane, academic discipline. We have forgotten the wisdom of a significant portion of this nation's founders, the Puritans, who understood that for theology to have meaning, it must permeate every aspect of life. In his essay "The Theological Task and Theological Method," Stephen Williams has eloquently articulated the failure of contemporary theology to address the theological aspects of daily living adequately.[24] The theological community must take up the issues we raise in this book and lift its sights from its own intradisciplinary conversations to an interdisciplinary engagement with medicine, philosophy, law, science, industry, and the lay community. For only in the context of a robust, practical theology of living can a workable theology of technology, and biotechnology, be developed. Our thoughts in this book can only serve as a starting point for the larger engagement that must occur and must occur soon. We acknowledge that none of the authors is a formally trained theologian, but each of us in our respective fields as healers, philosophers, and ethicists, in conversation with theologians, has been moved by the urgency to stimulate movement toward a theology of biotechnology.

The Formation of Communities of Responsibility and Accountability

Furthermore, we urge Christians to develop communities of technological responsibility and accountability. The external forces influencing the church and each of its members are significant, and in many cases far more influential than the church itself. Surveys have well documented the tragic degree of compromise and syncretism in the church in the United States.[25] Yet many of the choices we as Christians should make concerning technology are undoubtedly contra mundum, against the world and our cultural mandates of narcissism, materialism, and excessive consumption. Saying no to technicism, scientism, and the phoniness of our fashion-obsessed society will put us at odds with many of our culture's industrial and political bases, its media, and the entertainment industry. The embryonic stem cell issue illustrates that no level of prejudice, discrimination, censorship, name calling, antireligious bigotry, and academic dishonesty is beyond certain members of the press, academia, politics, and industry.

Having experienced some of this mistreatment ourselves, and having supported colleagues who have undergone similar and even worse mistreatment, we understand the pressures to conform or, worse, to remain silent. Our goal here is not to engage in naming victims or offenders, nor would it be appropriate for us to do so. Rather, our purpose is to highlight the desperate need for developing a faith community that will stand by, support, and protect when it can those who make choices honoring to God and consistent with His goals for our lives. We must have communities that are dedicated to genuine human flourishing, not just financial and social ease. This type of supportive community is what the body of Christ should be and must become.

Reformulation of Education and Critical Assessment Skills

We must approach education, particularly religious education, in an entirely different manner. Critical thinking skills are sadly lacking in our educational systems. We do not instill a healthy skepticism of our culture's technological conceit. Throughout the stem cell debate, for instance, outrageous claims of cures for practically everything have been bandied about and accepted uncritically by an undereducated and gullible populace. We have become either numb to or naively receptive to the claims of what amount to contemporary snake-oil salesmen instead of teaching ourselves and our children to challenge the hollow worldviews underlying emotional appeals to "Buy this now!" With itchy ears (2 Tim. 4:3), we respond as lambs, led by a false shepherd into the slaughterhouse of sales and misguided science, instead of treating their wares as the logical and factual farces they are. We do not test claims to prove whether they are indeed true (Acts 17:11). This also must change.

As each day passes, more and more innovations in biotechnology will be thrust upon humankind. Many will be welcomely received as they provide us with better tools to pursue the ministry of healing. Others, or perhaps some of the same, will also provide means to alter the human form and become tools for the remaking of human beings into something different. Many of these new technologies will be benign; others will pose risks to the environment. All will need to be weighed in terms of cost/benefit analysis, and the costs considered must include social, psychological, and spiritual as well as physical impact.

Conclusion: For the Human Good

Many will doubtless embrace whatever new technology comes along. Since the Fall, humans have sought to wrest control of their destiny from God. The physicist Freeman J. Dyson has asserted:

> The artificial improvement of human beings will come, one way or another, whether we like it or not, as soon as the progress of biological understanding makes it possible. When people are offered technical

> means to improve themselves and their children, no matter what they conceive improvement to mean, the offer will be accepted. Improvement may mean better health, longer life, a more cheerful disposition, a stronger heart, a smarter brain, the ability to earn more money as a rock star or baseball player or business executive. The technology of improvement may be hindered or delayed by regulation, but it cannot be permanently denied.[26]

Yet while there will be those who pursue this path, it is not inevitable that all will. We are all obligated to question each new development, each proposed technology, to ask whether it will assist or hinder our quest to be as Christ to the world. As a community of believers, Christians must be willing to say "no," even when unpopular, even if somehow it would seem that we may lose our competitiveness in one sphere or another. We must not be conformed to the pattern of the world but to the image of Christ (Rom. 6:4–14, 8:29, 12:2). We do this not just for ourselves but also for all in the family of humanity. The Gospel of Christ is the only road to true human flourishing, and we who claim Christ's name must witness to this reality. The more we compromise with the degraded view of humanity present in our world, especially in our culture, the more we decrease the joy and happiness that comes from genuine communion with God and one another. The theologian Henri J. M. Nouwen once observed:

> Beneath all the great accomplishments of our time there is a great current of despair. While efficiency and control are the great aspirations of our society, the loneliness, isolation, lack of friendship and intimacy, broken relationships, boredom, feelings of emptiness and depression, and a deep sense of uselessness fill the hearts of millions of people in our success-oriented world.[27]

No technology can resolve the pain and the emptiness in human hearts. Bearers and living witnesses of Christ's truth can, but we must be willing to be Christ's living body. This will require careful choices about technology. Mumford optimistically wrote in 1951:

> The ideas and ideals that will transform our civilization, restoring initiative to persons and delivering us from the more lethal operations of automatism, are already in existence. . . . Indeed the very persons who will

> make the critical decisions, when a singular moment presents itself, are already it seems probable, alive: it is even possible that a decisive change is already in operation, though as thoroughly hidden to us as the future of Christianity was to Pontius Pilate.[28]

Indeed, these ideals are present—they are the Gospel of Christ, the true path to human flourishing. And all who bear his name must make the critical decisions, for the singular moment is now. This is the time when, led by scripture and the Holy Spirit, the minds of all saints, scholar or layperson, must take up the challenge of deciding about technology.

Notes

Notes to Chapter 1

1. Andy Miah, *Genetically Modified Athletes: Biomedical Ethics, Gene Doping and Sport* (London: Routledge, 2004).

2. This is the title of the book by Carl Elliott, *Better Than Well: American Medicine Meets the American Dream* (New York: W. W. Norton, 2003).

3. American Society for Aesthetic Plastic Surgery, 2004 Statistics. http://www.surgery.org/press/statistics-2004.php.

4. See "Nootropic Research" online at the Immortality Institute, a nonprofit educational organization whose mission is "to conquer the blight of involuntary death," http://www.imminst.org/forum/index.php?s=&act=SF&f=199. Also see "Nootropic Agents," http://nootropics.com/sources.html; "Smart Drugs" FAQ, http://www.fiu.edu/~mizrachs/nootropics.html; and James South, "Nootropics: Reviewing Piracetam and Analogues," http://smart-drugs.net/ias-nootropics.htm.

5. See, e.g., Richard Lynn, *Eugenics: A Reassessment* (Westport, Conn.: Praeger, 2001); and Dorothy C. Wertz and John C. Fletcher, *Genetics and Ethics in Global Perspective* (Boston: Kluwer Academic Publishers, 2004).

6. For a sense of marketplace pressure for eugenics, see Philip Kitcher, *The Lives to Come: The Genetic Revolution and Human Possibilities* (New York: Simon & Schuster, 1996); Lee M. Silver, *Remaking Eden: How Genetic Engineering and Cloning Will Transform the American Family* (New York: Avon Books, 1998); and Gregory Stock, *Redesigning Humans: Our Inevitable Genetic Future* (Boston: Houghton Mifflin, 2002).

7. Works upholding the desirability of genetically reengineering the human species include Stock, *Redesigning Humans*; Silver, *Remaking Eden*; Lynn, *Eugenics*; Allen Buchanan, Dan W. Brock, Norman Daniels, and Daniel Wikler, *From Chance to Choice: Genetics and Justice* (New York: Cambridge University

Press, 2000); Gregory E. Pence, *Who's Afraid of Human Cloning?* (Lanham, Md.: Rowman & Littlefield, 1998); and Glenn McGee, *The Perfect Baby: A Pragmatic Approach to Genetics* (Lanham, Md.: Rowman & Littlefield, 1997).

8. For additional introductions to these fields, see C. Christopher Hook, "Techno sapiens: Nanotechnology, Cybernetics, Transhumanism and the Remaking of Humankind," in *Human Dignity in the Biotech Century: A Christian Vision for Public Policy*, ed. Charles W. Colson and Nigel M. de S. Cameron (Downers Grove, Ill.: InterVarsity Press, 2004); and C. Christopher Hook, "Cybernetics and Nanotechnology," in *Cutting Edge Bioethics: A Christian Exploration of Technologies and Trends*, ed. John F. Kilner, C. Christopher Hook, and Diann Uustal (Grand Rapids: Eerdmans, 2002).

9. Key early work by Norbert Weiner (1894–1964) includes *The Human Use of Human Beings: Cybernetics and Society* (1st ed., Boston: Houghton Mifflin, 1950; 2nd ed., New York: Da Capo Press, 1988), 16; and *Cybernetics: Or Control and Communication in the Animal and the Machine*, 1st ed. (Cambridge, Mass.: MIT Press, 1948).

10. W. Ross Ashby, *Introduction to Cybernetics* (London: Chapman & Hall, 1957), 1.

11. Manfred Clynes and Nathan S. Kline, "Cyborgs and Space," *Astronautics*, September 1960, 26–27, 74–75.

12. These applications of cybernetics in neuroscience have been reported by Martin Jenkner, Bernt Müller, and Peter Fromherz, "Interfacing a Silicon Chip to Pairs of Snail Neurons Connected by Electrical Synapses," *Biological Cybernetics* 84 (2001): 239–49; J. Wessberg, C. R. Stambaugh, J. D. Kralik, P. D. Beck, M. Laubach, J. K. Chapin, J. Kim, S. J. Biggs, M. A. Srinivasan, and M. A. Nicolelis, "Real-Time Prediction of Hand Trajectory by Ensembles of Cortical Neurons in Primates," *Nature* 408 (November 16, 2000): 1–65; and S. K. Talwar, S. Xu, E. S. Hawley, S. A. Weiss, K. A. Moxon, and J. K. Chapin, "Rat Navigation Guided by Remote Control," *Nature* 417 (May 2, 2002): 37–38.

13. The neuro-chip device is described by Peter Fromherz and Alfred Stett, "Silicon-Neuron Junction: Capacitive Stimulation of an Individual Neuron of a Silicon Chip," *Physical Review Letters* 75 (August 21, 1995): 1670–73; Stefano Vassanelli and Peter Fromherz, "Neurons from Rat Brain Coupled to Transistors," *Applied Physics* 65 (1997): 85–88; and in communications from Infineon Technologies AG, Munich (U.S. office in San Jose), a semiconductor system solutions company that is collaborating with the Max Planck Institute for Biochemistry in Tübingen (http://www.infineon.com/news/press/302_042e.htm). Recent reports from Infineon's website include "Neuro-Chip from Infineon Can Read Your Mind: New Findings in Brain Research Expected," press

release, Munich, February 11, 2003; and Steve Quayle, "Neuro-Chip: Soul Catcher Architecture Up and Running," February 21, 2003, http://stevequayle.com/News.alert/03_Genetic/030221.soul.catcher.chip.u.html.

14. Garrett B. Stanley, F. Li Fei, and Dan Yang, "Reconstruction of Natural Scenes from Ensemble Responses in the Lateral Geniculate Nucleus," *Journal of Neuroscience* 19 (September 11, 1999): 8036–42.

15. Reports on improvement of vision and devices for the blind include Stanley, Fei, and Yang, "Reconstruction of Natural Scenes"; Steven Kotler, "Vision Quest: A Half Century of Artificial-Sight Research Has Succeeded and Now This Blind Man Can See," *Wired*, September 2002, 95–101; and Victor D. Chase, "Mind over Muscles," *Technology Review*, March/April 2000, 39–45.

16. Kotler, "Vision Quest."

17. Chase, "Mind over Muscles."

18. Chicago's Optobionics Corporation describes its ASR® device—the Artificial Silicon Retina™—at http://www.optobionics.com, which links to a published paper reporting results for the first six patients: A. Y. Chow, V. Y. Chow, K. H. Packo, J. S. Pollack, G. A. Peyman, and R. Schuchard, "The Artificial Silicon Retina Microchip in the Treatment of Vision Loss from Retinitis Pigmentosa," *Archives of Ophthalmology* 122, no. 4 (April 2004): 460–69.

19. For progress in developing a memory chip, see Duncan Graham-Rowe, "World's First Brain Prosthesis Revealed," *New Scientist*, March 12, 2003, http://www.newscientist.com/news/news.jsp?id-ns99993488.

20. Ibid.

21. For opportunities and issues in nanotechnology, see Committee on Technology, Interagency Working Group on Neuroscience, Engineering, and Technology, National Science and Technology Council, *Nanotechnology: Shaping the World Atom by Atom*, December 1999, http://www.wtec.org/loyola/nano/IWGN.Public.Brochure/; National Nanotechnology Initiative website, http://www.nano.gov ; Foresight Institute (now also called the Foresight Nanotech Institute, with the logo "Advancing Beneficial Nanotechnology"), of Palo Alto, http://www.foresight.org.; and Chris Toumey, "Nanotech: Anticipating Public Reactions to Nanotechnology," *Techné—Research in Philosophy and Technology*, no. 2 (Winter 2004), http://scholar.lib.vt.edu/ejournals/SPT/v8n2/toumey.html#Bainbridge. Additional in-depth materials concerning nanotechnology include Charles P. Poole and Frank J. Owens, *Introduction to Nanotechnology* (Hoboken, N.J.: John Wiley & Sons, 2003); V. Balzani, A. Credi, and M. Venturi, *Molecular Devices and Machines: A Journey into the Nanoworld* (Weinheim, Germany: Wiley-VCH, 2003); and Liming Dai, *Intelligent Macromolecules for Smart Devices: From Materials Synthesis to Device Applications* (New York: Springer, 2004).

22. See Steven Glapa, *A Critical Investor's Guide to Nanotechnology*, 2002, and a press release of February 27, 2002, both at the In Realis website, http://www.inrealis.com. In Realis is a Silicon Valley consulting firm of which Glapa is president. Congress authorized funds for nanotechnology in H.R. 766, 108th Congress, 1st Sess., 2003; see http://www.house.gov/science/hearings/full03/may01/hr766.pdf. The Nanotechnology Research and Development Act of 2003 was signed by the president December 3, 2003, but at this writing has not yet been codified.

23. Richard Feynman, "There's Plenty of Room at the Bottom," address given at Annual Meeting of American Physical Society, California Institute of Technology, December 29, 1959; http://www.zyvex.com/nanotech/feyn man.html. First published in *Engineering and Science* (copyright holder), February 1960. Key works on nanotechnology's takeoff include Eric K. Drexler, *Engines of Creation: The Coming Era of Nanotechnology* (Garden City, N. Y.: Anchor Books/Doubleday, 1990 [1986, 1987]); Eric K. Drexler and Chris Peterson, *Unbounding the Future: The Nanotechnology Revolution* (New York: William Morrow, 1991); and Drexler, *Nanosystems: Molecular Machinery, Manufacturing, and Computation* (New York: Wiley Interscience, 1992).

24. On electronic miniaturization, see, e.g., Adrian Bachtold, Peter Hadley, Takeshi Nakanishi, and Cees Dekker, "Logic Circuits with Carbon Nanotube Transistors," *Science* 294 (November 9, 2001): 1317–20; and D. J. Hornbaker, S. K. Kahng, S. Misra, B. W. Smith, A. T. Johnson, E. J. Mele, D. E. Luzzi, and A. Yazdani, "Mapping the One-Dimensional Electronic States of Nanotube Peapod Structures," *Science* 295 (February 1, 2002): 828–31.

25. The Institute for Soldier Nanotechologies is part of the U.S. Army Research Office in the Department of Defense, http://www.aro.army.mil/sol diernano/.

26. Advances in targeting tumors are reported in the following: Michael McDevitt et al., "Tumor Therapy with Targeted Atomic Nanogenerators," *Science* 294 (November 16, 2001): 1537–40; Abraxane™ for injectable suspension, http://www.abraxane.com/; NCI [National Cancer Institute] Alliance for Nanotechnology in Cancer, brochure, *Cancer Nanotechnology: Going Small for Big Advances* (breast cancer), http://nano.cancer.gov/resource_center/cancer_nanotechnology_brochure.as p; communications of Introgen Therapeutics, corporate office in Austin, http://www.corporate-ir.net/ireye/ir_site.zhtml?ticker=ingn&script=2100; L. R. Hirsch et al., "Nanoshell-Mediated Near-Infrared Thermal Therapy of Tumors under Magnetic Resonance Guidance," *Proceedings of the National Academy of Sciences* 100 (November 11, 2003): 13549–54; pSivida Limited, a global nanotechnology company in Perth, Western Australia, which developed BioSilicon™, http://www.psivida

.com/default.asp; and Stephen C. Lee, Mark Ruegsegger, Philip D. Barnes, Bryan R. Smith, and Mauro Ferrari, "Therapeutic Nanodevices," in *Springer Handbook of Nanotechnology*, ed. Bharad Bhushan (New York: Springer, 2004).

27. Progress in nanomedicine in attacking infectious agents is reported in "Bacteria: Beware! Self-Assembled Cyclic Peptide Nanotubes Potentially Offer a Powerful Approach to Treating Bacterial Infections," Structure-Based Drug Design: A C&En Special Report, *Chemical & Engineering News* 79, no. 2 (August 6, 2001): 41–43, http://pubs.acs.org/cen/coverstory/7932/7932bacteria.html.

28. Reports on drug-delivery systems include Rainer H. Muller and Cornelia M. Keck, "Challenges and Solutions for the Delivery of Biotech Drugs: A Review of Drug Nanocrystal Technology and Lipid Nanoparticles," *Journal of Biotechnology* 113 (September 30, 2004): 151–70; R. H. Muller and C. M. Keck, "Nanoparticles for the Delivery of Genes and Drugs to Human Hepatocytes," *Nature Biotechnology* 21, no. 8 (August 21, 2003): 885–90; and Ezio Ricca and Simon M. Cutting, "Emerging Applications of Bacterial Spores in Nanobiotechnology," *Journal of Nanobiotechnology* 1 (December 15, 2003): 1–6.

29. For reports on diagnostic tools, see Park So-Jung, T. Andrew Taton, and Chad A. Mirkin, "Array-Based Electrical Detection of DNA with Nanoparticle Probes," *Science* 295, no. 5559 (February 22, 2002): 1503–6; James R. Heath, Michael E. Phelps, and Leroy Hood, "NanoSystems Biology, "*Molecular Imaging & Biology* 5, no. 5 (September–October 2003): 312–25; Amanda J. Haes, Lei Chang, William L. Klein, and Richard P. Van Duyne, "Detection of a Biomarker for Alzheimer's Disease from Synthetic and Clinical Samples Using a Nanoscale Optical Biosensor," *Journal of the American Chemical Society* 127, no. 7 (February 23, 2005): 2264–71; and Quantum Dot Corporation, Hayward, Calif. (Qdot[R] particles), at http://www.qdots.com/live/index.asp.

30. On blood replacement, see Robert A. Freitas Jr., "Respirocytes: A Mechanical Artificial Red Cell: Exploratory Design in Medical Nanotechnology," revision of 1996 paper, copyright now 1996–99, http://www.foresight.org/Nanomedicine/Respirocytes.html.

31. For information on improvements of prosthetic devices and implants, see Robert A. Freitas Jr., *Nanomedicine, Volume 1: Basic Capabilities* (Georgetown, Tex.: Landesbioscience, 1999); B. C. Crandell, ed., in *Nanotechnology: Molecular Speculations on Global Abundance* (Cambridge, Mass.: MIT Press, 1999); and BECON (NIH Bioengineering Consortium), *Nanoscience and Nanotechnology: Shaping Biomedical Technology: Symposium Report*, June 2000, http://grants.nih.gov/becon/becon_symposia.htm.

32. See Richard Crawford, "Cosmetic Nanosurgery," 61–80, in *Nanotechnology*, ed. Crandell.

33. For "bottom-up" advances in using biological materials operating like machines at the nanoscale, particularly work by the Montemagno group, see David S. Goodsell, *Bionanotechnology: Lessons from Nature* (Hoboken, N.J.: Wiley-Liss, 2004); Richard A. L. Jones, *Soft Machines: Nanotechnology and Life* (New York: Oxford University Press, 2004); Ralph S. Greco, Fritz B. Prinz, and R. Lane Smith, eds., *Nanoscale Technology in Biological Systems* (New York: CRC Press, 2005); Ricky K. Soong, George D. Bachand, Hercules P. Neves, Anatoli G. Olkhovets, Harold G. Craighead, and Carlo D. Montemagno, "Powering an Inorganic Nanodevice with a Biomolecular Motor," *Science* 290, no. 5496 (November 2000): 1555–58; Carlo D. Montemagno, George D. Bachand, Scott Stelick, and Marlene Bachand, "Constructing Biological Motor Powered Nanomechanical Devices," *Nanotechnology* 10 (1999): 225–31, http://www.foresight.org/Conferences/MNT6/Papers/Montemagno/; and Dan Shu and Peixuan Guo, "A Viral RNA That Binds ATP and Contains a Motif Similar to an ATP-Binding Aptamer from SELEX," *Journal of Biological Chemistry* 278 (February 28, 2003): 7119–25.

34. Montemagno et al., "Constructing Biological Motor Powered Nanomechanical Devices."

35. Shu and Guo, "Viral RNA."

36. A second bottom-up approach, using biomolecules for computation, is reported in Natasa Jonoska and Nadrian C. Seeman, eds., in *DNA Computing: 7th International Workshop on DNA-Based Computers* (Tampa, June 2001) (New York: Springer, 2002); Tanya Sienko, Andrew Adamatzky, Nicholas G. Rambidi, and Michael Conrad, *Molecular Computing* (Cambridge, Mass.: MIT Press, 2003); Yaakov Benenson, Tamar Paz-Elizur, Rivka Adar, Ehud Keinan, Zvi Livneh, and Ehud Shapiro, "Programmable and Autonomous Computing Machine Made of Biomolecules," *Nature* 414 (November 22, 2001): 430–34; Leonard M. Adleman, "Molecular Computation of Solutions to Combinatorial Problems," *Science* 266 (November 11, 1994): 1021–24; Dirk Faulhammer, Anthony R. Cukras, Richard J. Lipton, and Laura F. Landweber, "Molecular Computation: RNA Solutions to Chess Problems," *Proceedings of the National Academy of Sciences* 97 (February 2000): 1385–13; Yaakov Benenson, Rivka Adar, Tamar Paz-Elizur, Zvi Livneh, and Ehud Shapiro, "DNA Molecule Provides a Computing Machine with Both Fuel and Data," *Proceedings of the National Academy of Sciences* 100, no. 5 (March 4, 2003): 2191–96; and Nicholas Mano, Fei Mao, and Adam Heller, "A Miniature Biofuel Cell Operating in a Physiological Buffer," *Journal of the American Chemical Society* 124 (2002): 12962–63.

37. Faulkamen et al., "Molecular Computation."

38. Benenson et al., "DNA Molecule."

39. Mano, Mao, and Heller, "Miniature Biofuel Cell."

40. Progress in integrating the biological with the inorganic is reported in Chaunbin Mao, Christine E. Flynn, Andrew Hayhurst, et al., "Viral Assembly of Oriented Quantum Dot Nanowires," *Proceedings of the National Academy of Sciences* 100 (June 10, 2003): 6946–51; Clara S. Chan, Gelsomina DeStasio, Susan A. Welch, et al., "Microbial Polysaccharides Template Assembly of Nanocrystal Fibers," *Science* 303 (March 12, 2004): 1656–58; Kimberly Hamad-Schifferil, John J. Schwartz, Aaron T. Santos, Shuguang Zhang, and Joseph M. Jacobson, "Remote Electronic Control of DNA Hybridization Through Inductive Coupling to an Attached Metal Nanocrystal Antenna," *Nature* 415 (January 10, 2002): 152–55; H. Kobayashi, Mads Kaern, Michihiro Araki, et al., "Programmable Cells: Interfacing Natural and Engineered Gene Networks," *Proceedings of the National Academy of Sciences* 101 (June 1, 2004): 8414–19; and Martyn Amos, ed., *Cellular Computing* (New York: Oxford University Press, 2004).

41. Authors seeking to move ahead with "enhancement" now include Ray Kurzweil, *The Age of Spiritual Machines: When Computers Exceed Human Intelligence* (New York: Viking Press, 1999); Ray Kurzweil and Terry Grossman, *Fantastic Voyage: The Science behind Radical Life Extension—Live Long Enough to Live Forever* (Emmaus, Pa.: Rodale, 2004); Ramez Naam, *More Than Human: Embracing the Promise of Biological Enhancement* (New York: Broadway Books, 2005); Joel Garreau, *Radical Evolution: The Promise and Peril of Enhancing Our Minds, Our Bodies—and What It Means to Be Human* (New York: Doubleday, 2004); and James Hughes, *Citizen Cyborg: Why Democratic Societies Must Respond to the Redesigned Human of the Future* (Cambridge, Mass.: Westview Press, 2004).

42. For information on "transhumanism," see the World Transhumanist Association website, http://www.transhumanism.org; Nick Bostrom, "What Is Transhumanism?" http://www.nickbostrum.com; and Bostrom, "The Transhumanist FAQ," http://www.transhumanism.org/resources/faq.html.

43. The National Science Foundation published the first print edition of *Converging Technologies for Improving Human Performance*, by Mihail C. Roco and William Sims Bainbridge (North Arlington, Va.: National Science Foundation, 2002; Dordrecht and Boston: Kluwer, 2003); full text available at http://wtec.org/ConvergingTechnologies/.

44. Kevin Warwick, *I, Cyborg* (London: Century, 2002).

Notes to Chapter 2

1. Edward Tenner, *Why Things Bite Back: Technology and the Revenge of Unintended Consequences* (New York: Alfred A. Knopf, 1996).

2. Pertti J. Pelto, *The Snowmobile Revolution: Technology and Social Change in the Arctic* (Menlo Park, Calif.: Cummings Publishing Company, 1973), 152.

3. Ibid., 152.

4. David E. Nye, *America as Second Creation: Technology and Narratives of New Beginnings* (Cambridge, Mass.: MIT Press, 2003), 2.

5. Ibid., 13.

6. James Fenimore Cooper, *The Chainbearer* (New York: James G. Gregory 1864), 97.

7. Timothy Walker is cited in *The Machine in the Garden: Technology and the Pastoral Idea in America*, by Leo Marx (New York: Oxford University Press 1964), 182–83.

8. John Locke, *Second Treatise on Government*, ed. Thomas Peardon (Indianapolis: Bobbs-Merrill, 1952 [1690]), 23.

9. The notion of Lockean property rights has been very influential in American technological development, not least because it has been used to inform the system of patenting inventions. However, Lockean notions of property acquisition are not without their problems.

10. Ferguson is cited by Nye, *America as Second Creation*, 96.

11. Ibid., 148.

12. Ibid., 147.

13. Lewis Mumford, *Technics and Civilization* (New York: Harcourt, Brace, 1934).

14. Ibid., 109.

15. Ibid., 155.

16. Ibid., 353.

17. Ibid., 355.

18. Ibid., 254.

19. Nye, *America as Second Creation.*

20. Ibid., 295.

21. Ibid., 297.

22. Ibid., 298.

23. Ibid. Jim O'Brien, "A Beaver's Perspective on North American History," has been reprinted in *Major Problems in American Environmental History: Documents and Essays*, ed. Carolyn Merchant (Lexington, Mass: D. C. Heath, 1993), 78–83.

24. John Kenneth Galbraith, *The New Industrial State*, 2nd rev. ed (Boston: Houghton Mifflin, 1971), 12.

25. Steven V. Monsma, *Responsible Technology: A Christian Perspective* (Grand Rapids: Eerdmans, 1986), 19.

26. See Langdon Winner's penetrating critique, *Autonomous Technology: Technics-out-of-Control as a Theme in Political Thought* (Cambridge, Mass.: MIT Press, 1977).

27. David E. Nye, "Shaping Communication Networks: Telegraph, Telephone, Computer," *Social Research* 11 (September 22, 1997): 1067–91, at 1067.

28. Henry Petroski, *To Engineer Is Human: The Role of Failure in Successful Design* (New York: St. Martin's Press, 1985),

29. Graham R. Houston, *Virtual Morality: Christian Ethics in the Computer Age* (Leicester: Apollos, 1998), 69.

30. Monsma, *Responsible Technology*, 19.

31. C. S. Lewis, *The Abolition of Man* (New York: Collier Books, 1962 [1947]), 69.

32. Petroski, *To Engineer Is Human*, 2.

33. Kudzu, a native Japanese and Chinese plant that was transplanted in the United States (especially in the Southeast) originally to stop erosion, proved to be very hard to control because of its hardiness and fast growth. Sometimes growing several feet per day, kudzu has been known to cover entire homes or automobiles in a single season.

34. Monsma, *Responsible Technology*, 200.

35. Pelto, *Snowmobile Revolution*.

36. Ronald Cole-Turner, *The New Genesis: Theology and the Genetic Revolution* (Louisville: Westminster / John Knox Press, 1993), 108–9.

37. Of course I am alluding to Richard Dawkins, *The Blind Watchmaker: Why the Evidence of Evolution Reveals a Universe without Design* (New York: W. W. Norton, 1996).

38. Kevin Warwick, *I, Cyborg* (London: Century, 2002). On Warwick's implanted chip, see http://www.kevinwarwick.com.

39. Peter McGrath, "Building a Better Human," *Newsweek*, Special Edition, 2001, 46–49, at 46.

40. Note that the Islamic worldview also supports this special stewardship relationship. Cf. John L. Esposito, *Islam: The Straight Path*, 3rd ed. (New York: Oxford University Press, 1998), esp. 25–30, observing that "the essence of human uniqueness lies in one's vocation as God's representative on earth," given "as a divine trust" and deriving from the root Islamic (Quranic) principle "that the earth belongs ultimately to God and that human beings are its caretakers."

41. Bill Joy, "Why the Future Doesn't Need Us," *Wired*, April 2000, 1–11.

Notes to Chapter 3

1. William Kristol and Eric Cohen, *The Future Is Now: America Confronts the New Genetics* (Lanham, Md.: Rowman & Littlefield, 2002).

2. Ibid., xvii; Kristol and Cohen are referring to the title of the book by C. S. Lewis, *The Abolition of Man* (New York: Collier Books, 1962 [1947]).

3. Jeremy Rifkin, *The Biotech Century: Harnessing the Gene and Remaking the World* (New York: Penguin Putnam, 1998).

4. Ibid., 198. Rifkin further suggests that worldviews (cosmologies, in his term) provide a mirror of the day-to-day activity of a civilization (p. 201).

5. Ibid., 207.

6. Gregory Stock, *Redesigning Humans: Our Inevitable Genetic Future* (Boston: Houghton Mifflin, 2002), 44.

7. It should be noted that all worldview commitments are ultimately faith commitments, none of which can be scientifically verified. No worldview commitment should be discounted simply because it is grounded in a particular religious faith. The naturalist and the environmentalist are similarly making faith commitments in forming their assumptions. Everyone in this debate begins with a set of presuppositions.

8. Some naturalists claim that whatever exists either must itself be physical or must depend on or supervene upon the physical. Those who adopt the latter allow for emergent properties that are not themselves physical but that depend on the physical. E.g., the property of wetness emerges when hydrogen and oxygen come together in the right proportions. That various properties, such as mental properties, are not physical is true. But that they are emergent properties that depend solely on matter is questionable. Indeed, many—most likely, most—naturalists reject emergent properties in favor of the strict physicalism addressed here because, among other things, if mental properties are granted as emergent, then most likely the best explanation for their appearance is a theistic one. As many naturalists see, such emergent mental properties provide grounds for inferring the existence of God, so they reject emergent mental properties for this reason. For more on this, see J. P. Moreland, "The Argument from Consciousness," in *The Rationality of Theism*, ed. Paul Copan and Paul Moser (London: Routledge, 2003), 204–20.

9. These are spelled out in more detail in James Sire, *Discipleship of the Mind* (Downers Grove, Ill.: InterVarsity Press, 1990), 11–12.

10. E. O. Wilson, "The Biological Basis for Morality," *The Atlantic*, April 1998, 98–107.

11. Paul M. Churchland, *Matter and Consciousness: Contemporary Introduction to the Philosophy of Mind* (Cambridge, Mass.: MIT Press, 1984 [1st ed.; rev. ed., 1988.]), 21.

12. Evelyne Shuster, "Determinism and Reductionism: A Greater Threat Because of the Human Genome Project?" in *Gene Mapping: Using Law and Ethics as Guides*, ed. George J. Annas and Sherman Elias (New York: Oxford University Press, 1992), 120–31.

13. The views of the genome project community are well documented by Dorothy Nelkin and Susan M. Lindee, *The DNA Mystique: The Gene as a Cultural Icon* (New York: W. H. Freeman, 1995); the quotation here is on 41.

14. See Walter Gilbert, "A Vision of the Grail," in *The Code of Codes: Scientific and Social Issues in the Human Genome Project*, ed. Daniel J., Kevles and Leroy Hood (Cambridge, Mass.: Harvard University Press, 1992), 83; also cited in "The Gene Hunt," by Leon Jaroff, *Time*, March 20, 1989, 63–67, at 63. The commandment "know thyself" is cited in "DNA Blueprints, Personhood and Genetic Privacy," by Hugh Miller III, *Case Western Reserve University Health Matrix: Journal of Law-Medicine* 8, no. 2 (Summer 1998): 179–221, at 180.

15. Gilbert, "Vision of the Grail," 96.

16. Gilbert is quoted in "The Science of Metamorphoses," by Richard C. Lewontin, *New York Review of Books*, April 27, 1989, 18.

17. Francis Crick, *The Astonishing Hypothesis: The Scientific Search for the Soul* (New York: Scribner, 1994), 3.

18. James D. Watson, "The Human Genome Project: Past, Present, and Future," *Science* 248, no. 4951 (1990): 44–49, at 44.

19. Watson is quoted in Jaroff, "Gene Hunt," 67.

20. This is summarized from a more detailed account by Shuster, "Determinism and Reductionism," 120–21.

21. Howard L. Kaye, *The Social Meaning of Modern Biology: From Social Darwinism to Sociobiology* (New Haven, Conn.: Yale University Press, 1986), 55.

22. In a memorial statement about Robert Haynes (died December 22, 1998), the Sixteenth International Congress of Genetics in Toronto in 1988 is described as the largest congress ever held in the history of genetics research (online excerpt from B. A. Kuntz and R. C. Hanewald, *Environmental & Molecular Mutagenesis* 33: 257–65 [1999]; http://www.ems.us.org/who_we_are/memorial/bobhaynes.asp). For the proceedings and abstracts, see Robert H. Haynes and David B. Walden [16th International Congress of Genetics, Toronto, 1988] *Genome* 31 (1989), and Abstracts, *Genome* 30, Suppl. 1 (1988); and David E. Walden, "Report of the Secretary-General of the 16th International Congress of Genetics," *Genome* 31 (1989): 1127–38. Haynes's explicit statement about the replacement of the traditional Judeo-Christian view of hummanity with the advances in genetic manipulation has been picked

up by several commentators, available online. E.g., see an online essay by Jim Leffel at the Xenos Center, "Engineering Life: Defining Humanity in a Postmodern World" (http://www.xenos.org/essays/medeng.htm#end10), referencing in his note 10 Andrew Kimbrell, *The Human Body Shop: The Engineering and Marketing of Life* (San Francisco: HarperSanFrancisco, 1993), 233–34.

23. A prime example of a prominent molecular biologist who does not share this materialistic view of the world is the current director of the HGP, Francis S. Collins, M.D., Ph.D., who is a Christian. Collins has also suggested that there might be more theism in the molecular biology community than one might imagine, taking the form of theistic evolution or evolutionary theism. He also put the HGP in proper perspective when he suggested to the PBS television interviewer Bob Abernethy that "it [the human genome] is, after all, a set of instructions; but it does not tell us what being human is—nor, would I argue, it ever will" (week of October 8, 2004); http://www.pbs.org/wnet/religion_and_ethics/transcripts/collins.html.

24. Kevin Kelly, *Out of Control: The Rise of Neo-Biological Civilization* (New York: William Patrick Books, 1994), 14. Kelly's title merits comment. Of course, the discontinuity in the two processes is that natural evolution is conceived on a materialistic view of the world, apart from any intelligent design. Artificial evolution is anything but analogous to random naturalistic evolution because it is intelligent human beings who are engineering artificial "evolution."

25. Ibid., 55. Notice that Kelly's underlying view of a person is expressed in mechanistic terms.

26. Ibid., 14.

27. Ray Kurzweil, *The Age of Spiritual Machines: When Computers Exceed Human Intelligence* (New York: Viking Press, 1999).

28. Ibid., 3.

29. Kelly, *Out of Control*, 17.

30. Katharine Hayles, "How We Became Posthuman: Virtual Bodies," in *Cybernetics, Literature and Informatics* (Chicago: University of Chicago Press, 1999), 3.

31. Ibid., 11.

32. Damien Broderick, *The Spike: How Our Lives Are Being Transformed by Rapidly Advancing Technologies* (New York: Tom Doherty Associates, 2001), 205.

33. Ibid., 190.

34. Ibid., 191.

35. Kurzweil, *Age of Spiritual Machines*, 197. Interestingly, Kurzweil admits that there is a philosophical problem with consciousness arising from a strictly

material source. He is clearly uncomfortable with the idea of the "machine reflecting on itself." In addition, he admits that the problem of free will is complex. But it is clear that he admits to nothing beyond the material to explain these phenomena.

36. Ibid., 201, 203. Of course, what is not clear is how Kurzweil can have categories for the "spiritual" given his commitment to a naturalistic view of the person.

37. Ray Kurzweil, *The Singularity Is Near: When Humans Transcend Biology* (New York: Viking Press, 2005), 9.

38. See J. P. Moreland, "Timothy O'Connor and the Harmony Thesis: A Critique," *Metaphysica* 3, no. 2 (2002): 5–40.

39. J. L. Mackie, *Ethics: Inventing Right and Wrong* (New York: Penguin Books, 1977), 103. This same criticism of moral properties can also apply to other nonphysical properties, such as consciousness, mental states, secondary qualities, and the notion of mind.

40. For a general critique of genetic engineering applied to foods, animals, and human beings, see the set of essays in *Redesigning Life? The Worldwide Challenge to Genetic Engineering*, ed. Brian Tokar (London: Zed Books, 2001).

41. E.g., see the provocative work *Biopiracy: The Plunder of Nature and Knowledge*, by Vandana Shiva (Boston: South End Press, 1997), 24–30. She outlines a view she calls "the reductionist paradigm of biology," reflecting our discussion of genetic reductionism.

42. See this suggestion, e.g., in "Environmentalism: The Triumph of Politics," by Doug Bandow, *The Freeman*, September 1993, 32–39; cited in *Beyond Integrity: A Judeo-Christian Approach to Business Ethics*, 2nd ed., by Scott B. Rae and Kenman L. Wong (Grand Rapids: Zondervan 2004 [1996]).

43. Marc Lappe and Britt Bailey, *Against the Grain: Biotechnology and the Corporate Takeover of Your Food* (Monroe, Me.: Common Courage Press, 1998), 17.

44. Sheldon Krimsky and Roger Wrubel, *Agricultural Biotechnology and the Environment: Science, Policy and Moral Issues* (Urbana, Ill.: University of Illinois Press, 1996), 207.

45. Ibid., 208.

46. Shiva, *Biopiracy*, 123.

47. Martin Teitel and Kimberly A. Wilson, *Genetically Altered Food: Changing the Nature of Nature* (Rochester, Vt.: Park Street Press, 1999), 113.

48. Ibid., 112.

49. Ibid., 111.

50. David Ehrenfeld, "Hard Times for Diversity," in *Fatal Harvest: The Tragedy of Industrial Agriculture*, ed. Andrew Kimbrell (Washington, D.C.: Island Press, 2002), 110–18, at 111.

51. E.g., see Thomas Sieger Derr, "The Challenge of Biocentrism," in *Creation at Risk: Religion, Science and Environmentalism*, ed. Martin Cromartie (Grand Rapids: Eerdmans, 1995).

52. W. Michael Hoffman, "Business and Environmental Ethics," *Business Ethics Quarterly* 1, no. 2 (1991): 169–84; the citation here is on 173. Hoffman's suggestion is cited in *Beyond Integrity: A Judeo-Christian Approach to Business Ethics*, 2nd ed., by Scott B. Rae and Kenman L. Wong (Grand Rapids: Zondervan 2004 [1996]), 479.

53. See Peter Singer and Helga Kuhse, *Unsanctifying Human Life* (Malden, Mass.: Blackwell Publications, 2002); and Peter Singer, *Writings on an Ethical Life* (New York: HarperCollins, 2001). The term *speciesism* is used commonly throughout Singer's works attacking the traditional notion of the sanctity of life. For support of this traditional notion, see Tim Chappell, "In Defense of Speciesism," in *Human Lives: Critical Essays on Consequentialist Bioethics*, ed. David S. Odenberg and Jacqueline A. Laing (London: Macmillan, 1997). The term was coined by British psychologist Richard D. Ryder in 1970 (and used for the first time in a small historical and tourist brochure for the city of Oxford). Recently, in the August 6, 2005, *Guardian* ("All Beings That Feel Pain Deserve Human Rights"), Ryder recollects how the term first came to him "35 years ago" as he was taking his bath in Oxford. He terms the criterion for being included in this rights group *painience.*

54. For a recent and thorough philosophical and theological exposition of a Christian worldview, see J. P. Moreland and William Lane Craig, *Philosophical Foundations for a Christian World View* (Downers Grove, Ill.: InterVarsity Press, 2003).

55. An epistemology consistent with a Christian worldview would hold to what is called a correspondence view of truth. I.e., what we can know actually corresponds to what is "out there." It requires more than simply internal coherence to meet the test of truth. Though every person has lenses through which he or she perceives knowledge, it cannot be reduced to a social construction. For further discussion of epistemology from a Christian worldview, see Moreland and Craig, *Philosophical Foundations.*

56. For more expanded discussion of human dignity grounded in the image of God, see chapters 3 and 5 of the present volume. There is some concern expressed by adherents of biotechnology that it will lead to widening of existing inequalities and access to technological resources and that this constitutes a problem with the emerging biotechnologies, particularly in genetics. Interestingly, from a naturalist view of the world, it is not clear on what basis such a concern for equality is grounded. It would seem that these proponents are borrowing from Christian theism for their notion of dignity and equality. See

Lee M. Silver, *Remaking Eden: Cloning and Beyond in a Brave New World* (New York: Avon Books, 1997), 4–7, 240–50.

Notes to Chapter 4

1. Elements of the description of the various approaches to human dignity—as opposed to the critical analysis of them—have been developed from John F. Kilner's essay "Human Dignity," in *Encyclopedia of Bioethics*, 3rd ed. (New York: Macmillan), 1193–1200.

2. For more on the German use of the term, see Günter Dürig, "Der Grundrechtssatz von der Menschenwürde," *Archiv des Offentlichen Rechts* 81, no. 2 (1956): 117–57. Cf. Martin Hailer and Dietrich Ritschl, "The General Notion of Human Dignity and the Specific Arguments in Medical Ethics," in *Sanctity of Life and Human Dignity*, ed. Kurt Bayertz (Dordrecht: Kluwer, 1996); and Monika Burkart, *Das Recht, in Würde zu Sterben, ein Menschenrecht: eine Verfassungsrechtliche Studie zur Frage der Menschenwürdigen Grenze zwischen Leben un Tod* (Zurich: Schulthess, 1983).

3. United Nations, International Bill of Rights, 1966, Incorporating the Universal Declaration of Human Rights, 1948, with the International Covenant on Economic, Social, and Cultural Rights and the International Covenant on Civil and Political Rights. Entered into force January 3, 1976. Background reported in Office of the UN High Commissioner for Human Rights, Fact Sheet 2, June 1996; http://www.unhchr.ch/the menu6/2/Ps2.htm.

4. Council of Europe, Convention for the Protection of Human Rights and Dignity of the Human Being with Regard to the Application of Biology and Medicine: Convention on Human Rights and Biomedicine, No. 164, Strasbourg, Council of Europe, 1996. For more on the international use of the term, see Kurt Bayertz, "Human Dignity: Philosophical Origin and Scientific Erosion of an Idea," in *Sanctity of Life and Human Dignity* (Dordrecht: Kluwer, 1996), 73–90; cf. his introduction, xi–xix.

5. On the variable sense of dignity, see Herbert Spiegelberg, "Human Dignity: A Challenge to Contemporary Philosophy," in *Human Dignity: This Century and the Next*, ed. Rubin Gotesky and Ervin Laszlo (New York: Gordon and Breach, 1970), 39–64.

6. Peter Berger elaborates the difference between this variable sense of dignity as "honor" and what he identifies as the primary sense of dignity as a special status attributable to all people in Berger, "On the Obsolescence of the Concept of Honor," *European Journal of Sociology* 11 (1983): 339–47.

7. This notion that everything including human dignity can be placed on the same scale of value, though a claim about ethics, is often tied to the economic concept that everything has a cost that can be measured in common

(generally monetary) units. E.g., see Thomas Eger, Bernhard Nagel, and Peter Weise, "Effizienz und Menschenwürde—ein Gegensatz?" in *Okonomische Probleme des Zivilrechts*, ed, Claus Ott and Hans-Bernd Schafer (Heidelberg: Springer, 1991), 18–34.

8. A critical discussion of the implications of the Freudian school of thought for human dignity may be found in Willard Gaylin, "In Defense of the Dignity of Being Human," *Hastings Center Report* 14 (1984): 18–22.

9. The classic statement of this view is B. F. Skinner, *Beyond Freedom and Dignity* (New York: Alfred A. Knopf, 1971). It is critiqued in Francis Schaeffer, *Back to Freedom and Dignity* (London: Hodder & Stoughton, 1973).

10. E. O. Wilson's outlook is developed in Wilson, *Sociobiology: The New Synthesis* (Cambridge, Mass.: Harvard University Press, 1975). It is critiqued in Marie Augusta Neal, *The Just Demands of the Poor* (New York: Paulist, 1987).

11. E.g., see Richard Rorty, *Contingency, Irony, and Solidarity* (Cambridge: Cambridge University Press, 1989). Kevin Vanhoozer examines the limitations of this approach—including Michel Foucault's promotion of "the death of Man"—in Vanhoozer, "Human Being, Individual and Social," in *The Cambridge Companion to Christian Doctrine*, ed. Colin E. Gunton (Cambridge: Cambridge University Press, 1997), 158–88.

12. Ray Kurzweil, *The Age of Spiritual Machines: When Computers Exceed Human Intelligence* (New York: Viking Press, 1999).

13. Rorty, *Contingency*, 190.

14. Daryl Pullman, "Universalism, Particularism and the Ethics of Dignity," *Christian Bioethics* 7, no. 3 (2001): 333–58.

15. See Ruth Macklin, "Dignity Is a Useless Concept" (editorial), and the follow-up correspondence in the *British Medical Journal* 327, no. 7429 (December 20, 2003): 1419–20.

16. E.g., see Joseph F. Fletcher, *Humanhood: Essays in Biomedical Ethics* (Buffalo: Prometheus Books, 1979).

17. Marcus Aurelius, *Meditations*, trans. Gregory Hays (New York: Random House Modern Library, 2003 [167]); Giovanni Pico della Mirandola, "Oration on the Dignity of Man," in *The Renaissance Philosophy of Man*, ed. Ernst Cassirer, Paul Oskar Kristeller, and John Herman Randall Jr. (Chicago: University of Chicago Press, 1956 [1486]); John Locke, *An Essay Concerning Human Understanding*, ed. Peter Nidditch (Oxford: Clarendon, 1989 [1690]); Blaise Pascal, *Thoughts: Blaise Pascal*, ed. C. W. Eliot and trans. W. F. Trotter, Harvard Classics, vol. 18 (New York: P. F. Collier & Sons, 1910 [1660]), fragment 347.

18. Immanuel Kant, *Groundwork of the Metaphysics of Morals*, trans. H. J. Paton (New York: Harper & Row, 1964 [1785]), 102.

19. Ibid., 107, 103.

20. Thomas E. Hill, *Dignity and Practical Reason in Kant's Moral Theory* (Ithaca, N.Y.: Cornell University Press, 1992).

21. Kant, *Groundwork*, 100.

22. Ibid., 108.

23. Ibid., 105.

24. Ibid.

25. Deryck Beyleveld and Roger Brownsword, *Human Dignity in Bioethics and Biolaw* (New York: Oxford University Press, 2001).

26. Alan Gewirth, *Reason and Morality* (Chicago: University of Chicago Press, 1978).

27. Beyleveld and Brownsword, *Human Dignity*, 5.

28. For a review of communitarian accounts of dignity, see ibid., 26, 65.

29. Gewirth, *Reason*, 121–22, 141.

30. Beyleveld and Brownsword, *Human Dignity*, 112–13.

31. For a fuller demonstration of how Kant's own position leads to this conclusion, see R. W. Krutzen, "Intrinsic Value and Human Dignity," *Religious Studies and Theology* 15 (1996): 35–47.

32. Leon R. Kass, *Life, Liberty and the Pursuit of Dignity: The Challenge for Bioethics* (San Francisco: Encounter Books, 2002), 117–18. Francis Fukuyama makes a similar point, especially on the important place of emotions, in Fukuyama, *Our Posthuman Future: Consequences of the Biotechnology Revolution* (New York: Farrar, Straus & Giroux, 2002).

33. Kass, *Life*, 6, 9.

34. Pepita Haezrahi, "The Concept of Man as End-in-Himself," in *Kant: Foundations of the Metaphysics of Morals*, ed. Robert P. Wolff and trans. Lewis White Beck (Indianapolis: Bobbs-Merrill 1982), 317.

35. Hailer and Ritschl, "General Notion," 99.

36. UNESCO, *Universal Declaration on the Human Genome and Human Rights* (Paris: UNESCO, 1997).

37. Fukuyama, *Our Posthuman Future*, 170.

38. E.g., see Tetsuya Inoue, "Dignity of Life," in *Human Dignity and Medicine, Proceedings of the Fukui Bioethics Seminar, Japan*, ed. J. Bernard, K. Kajikawa, and N. Fujiki (Amsterdam: Elsevier 1988), 11–16; cf. James S. Dalton, "Human Dignity, Human Rights, and Ecology: Christian, Buddhist, and Native American Perspectives," in *Made in God's Image: The Catholic Vision of Human Dignity*, ed. Regis A. Duffy and Angelus Gambatese (New York: Paulist Press 1999), 29–53.

39. E.g.: For Judaism, see Haim H. Cohn, "On the Meaning of Human Dignity," *Israel Yearbook on Human Rights* 13 (1983): 226–51. For Christianity, see Jurgen Moltmann, *On Human Dignity: Political Theology and Ethics*

(Philadelphia: Fortress Press, 1984). For Islam, see Heiner Bielefeldt, Winfred Brugger, and Klaus Dicke, eds., *Würde und Recht des Menschen* (Würzburg: Königshausen & Nevmann, 1992). Cf. the view of human beings as God's representatives in *Islam: The Straight Path*, 3rd ed., by John L. Esposito (New York: Oxford University Press 1998), 25–26, citing several passages in the Quran.

40. E.g., see Bayertz, "Human Dignity." For more reviews of the dignity concept in this long-standing public tradition drawing on religious traditions, see Ronald B. Allen, *The Majesty of Man: The Dignity of Being Human* (Grand Rapids: Kregel, 2000); Garth Baker-Fletcher, *Somebodyness: Martin Luther King and the Theory of Dignity* (Minneapolis: Fortress Press, 1993); Miroslav Bednar, ed., *Human Dignity: Values and Justice* (Washington, D.C.: Council for Research in Values and Philosophy, 1999); Jolana Poláková, "The Struggle for Human Dignity in Extreme Situations," in *Human Dignity*, ed. Bednar, chapter 2, http://www.crvp.org/book/series04/IVA-18/chapter_ii.htm; Ernst Bloch, *Natural Law and Human Dignity*, trans. Dennis J. Schmidt (Cambridge, Mass.: MIT Press, 1986); Goran Collste, *Is Human Life Special: Religious and Philosophical Perspectives on the Principle of Human Dignity* (New York: Peter Lang, 2002); Charles W. Colson and Nigel M. de S. Cameron, eds., *Human Dignity in the Biotech Century: A Christian Vision for Public Policy* (Downers Grove, Ill.: InterVarsity Press, 2004); Geoffrey G. Drutchas, *Is Life Sacred?* (Cleveland: Pilgrim Press, 1998); D. A. du Toit, "Anthropology and Bioethics," *Ethics and Medicine* 10, no. 2 (Summer 1994): 35–42; Dan Egonsson, *Dimensions of Dignity: The Moral Importance of Being Human* (Dordrecht: Kluwer, 1998); Jürgen Habermas, *The Future of Human Nature*, trans. Max Pensky and William Rehq (Cambridge: Polity, 2003); Mats G. Hansson, *Human Dignity and Animal Well-Being: A Kantian Contribution to Biomedical Ethics* (Stockholm: Almqvist & Wiksell, 1991); Edwin C. Hui, *At the Beginning of Life: Dilemmas in Theological Bioethics* (Downers Grove, Ill.: InterVarsity Press, 2002); John Kavanaugh, *Who Counts As Persons? Human Identity and the Ethics of Killing* (Washington, D.C.: Georgetown University Press, 2001); John F. Kilner, Arlene B. Miller, and Edmund D. Pellegrino, eds., *Dignity and Dying* (Grand Rapids: Eerdmans, 1996); Bartha M. Knoppers, *Human Dignity and Genetic Heritage* (Ottawa: Law Reform Commission of Canada, 1991); Aurel Kolnai, "Dignity," *Journal of Philosophy* 51 (1976): 251–71; Krutzen, "Intrinsic Value"; Karen Lebacqz, "On the Elusive Nature of Respect," in *The Human Embryonic Stem Cell Debate*, ed. Suzanne Holland, Karen Lebacqz, and Laurie Zoloth (Cambridge, Mass.: MIT Press, 2001), 149–62; Ted Peters, "Embryonic Stem Cells and the Theology of Dignity," in *Human Embryonic Stem Cell Debate*, ed. Holland et al., 127–40; Michael J.

Meyer and William A. Parent, eds., *The Constitution of Rights: Human Dignity and American Values* (Ithaca, N.Y.: Cornell University Press, 1992); Leslie F. Moser, *The Struggle for Human Dignity* (Los Angeles: Nash Publishing, 1973); Michael S. Pritchard, "Human Dignity and Justice," *Ethics* 82 (1972): 106–13; Oscar Schachter, "Human Dignity as a Normative Concept," *American Journal of International Law* 77 (1983): 848–54; and Michael A. Smith, *Human Dignity and the Common Good in the Aristotelian-Thomistic Tradition* (Lewiston, N.Y.: Mellen Press, 1995).

41. For a historical overview, see G. C. Berkouwer, *Man: The Image of God* (Grand Rapids: Eerdmans, 1975).

42. See Gerald Bray, "The Significance of God's Image in Man," *Tyndale Bulletin* 42, no. 2 (1991): 195–225.

43. Claus Westermann, *Genesis 1–11*, trans. John J. Scullion (Minneapolis: Augsburg Press, 1984).

44. Illustrations of the most influential figures involved may be found in Millard Erickson, *Christian Theology* (Grand Rapids: Baker, 1985).

45. This notion is developed later in the chapter in conjunction with the discussion of "speciesism."

46. This idea is developed further by Karl Barth, *Church Dogmatics* (Edinburgh: T. & T. Clark, 1958), vol. 3, 2.

47. "Innocent" in this sentence refers to people who are not guilty of participating in an action such as unjust aggression in war or first-degree murder that would warrant the forfeiture of their lives, if there are such actions.

48. This central biblical notion, as variously nuanced by Rudolf Bultmann, *Theologie des Neuen Testaments* (Tübingen: J. C. B. Mohr, 1953), 38, and other biblical scholars, is critically assessed and applied to bioethics in John F. Kilner, *Life on the Line: Ethics, Aging, Ending Patients' Lives, and Allocating Vital Resources* (Grand Rapids: Eerdmans 1990), 22ff.

49. For more on the relation between being a neighbor and being an image of God, see Donal O'Mathuna, "The Bible and Abortion: What of the 'Image of God'?" in *Bioethics and the Future of Medicine*, ed. John F. Kilner, Nigel M. de S. Cameron, and David Schiedermayer (Grand Rapids: Eerdmans, 1995), 205–6.

50. E.g., see William P. Cheshire, "Toward a Common Language of Human Dignity," *Ethics and Medicine* 18, no. 2 (2000): 7–10.

51. Such critics include Peter Singer, e.g., in *Rethinking Life and Death* (Oxford: Oxford University Press, 1995); Helga Kuhse, e.g., in "Is There a Tension between Autonomy and Dignity?" in *Bioethics and Biolaw*, vol. 2, *Four Ethical Principles*, ed. Peter Kemp, Jacob Rendtorff, and N. Mattson Johansen (Copenhagen: Rhodos International Science and Art Publishers and Centre for

Ethics and Law, 2000), 61–74. This criticism has been countered by Tim Chappell, "In Defense of Speciesism," in *Human Lives: Critical Essays on Consequentialist Bioethics*, ed. David S. Odenberg and Jacqueline A. Laing (London: Macmillan, 1997), 96–108.

52. Once the courts had allowed euthanasia to take place without prosecution for an extended period of years, the Dutch Committee to Investigate the Medical Practice Concerning Euthanasia released official annual figures documenting the practice of euthanasia in The Netherlands. The report documents nearly 6,000 cases in which actions were taken—without patients' permission—with the intent to end patients' lives (i.e., cases of direct killing or provision of morphine in excessive doses with an intent to end life). In nearly one-fourth of those cases, the patients were still mentally competent. See Richard Fenigsen, "The Report of the Dutch Governmental Committee on Euthanasia," *Issues in Law and Medicine* 7 (Winter 1991): 339–44. For a recent medical consideration of dignity in care at the end of life, see Harvey M. Chochinov, "Dignity-Conserving Care: A New Model for Palliative Care—Helping the Patient Feel Valued," *Journal of the American Medical Association* 287, no. 7 (May 1, 2002): 2253–60.

53. Krutzen, "Intrinsic Value," 38.

Notes to Chapter 5

1. The United States is clearly our focus, although much if not all of what we say is applicable to the developed countries or, in John Paul II's terms, the "civilization of consumption" of the West (Papal Encyclical *Sollicitudo res socialis*, December 30, 1987, sec. 28).

2. Dietrich Bonhoeffer, *Creation and Fall: A Theological Exposition of Genesis 1–3* (Minneapolis: Fortress Press, 1997). This is volume 3 of *Dietrich Bonhoeffer's Works*, now in progress.

3. Robin W. Lovin, *Reinhold Niebuhr and Christian Realism* (Cambridge: Cambridge University Press, 1995). This is not the time or the place to unpack ethical naturalism and moral realism. Suffice it to say that Lovin is committed to the view that there is a "there there," that there are truths to be discerned about the world and that the world is not just so much putty in our conceptually deft hands. The world exists independent of our minds, but our minds possess the wonderful capacity to apprehend the world—up to a point, given the fallibility of reason.

4. Judith Jarvis Thomson, "A Defense of Abortion," *Journal of Philosophy and Public Affairs* 1, no. 1 (Fall 1971): 47–66. Thomson, known for her current support of physician-assisted suicide, first made her reputation in this first

issue of the *Journal of Philosophy and Public Affairs* by providing justifications for abortion by analogizing from a woman hooked up during her sleep to a violinist for whom she was then required to provide life support, to a woman in relationship to the fetus she is carrying. Thomson claimed that the woman would be within her rights to unhook the violinist, even if it meant his or her death; similarly, a woman is not required to carry a fetus to term. I have never understood why a reasonable person would find this argument compelling. Fetuses do not get attached covertly but emerge as a result of action in which the woman is implicated. As well, the fetus's dependence on the mother for sustenance for nine months is part of the order of nature—it simply is the way humans reproduce. There are many ways to sustain violinists in need of life support, and an adult violinist is scarcely analogous in any way to the life of a human being in situ.

5. Lovin, *Reinhold Niebuhr*, 123.

6. Ibid., 126.

7. Ibid., 94.

8. Ibid., 130.

9. This, too, is more complex than simply acquiescence. E.g., on the matter of abortion, there is enormous popular support for some forms of restriction and restraint. The elite culture (the media, those with incomes more than $50,000 per year, lawyers, as the most reliable social science studies demonstrate) long ago fell into lockstep with an absolute abortion "right," including partial-birth abortion (dilatation and extraction), a practice found by surveys to be abhorrent to most Americans. So on the level of opinion, not all is homogeneous. But this opinion rarely translates into action of any sort. Thus the atrophy of civic habits of the past four decades or so goes hand in hand with the triumph of projects that constitute flights from finitude.

10. This situation is not ours alone, of course, but here we will concentrate primarily on North American culture in depicting this obsession and grappling with its hold on the collective psyche.

11. Charles Taylor, *The Ethics of Authenticity* (Cambridge, Mass.: Harvard University Press, 1992); orig. pub. in Canada as *The Malaise of Modernity* (Toronto: Anansi, 1991). Just to be clear at the outset, this is not intended to issue strictures against any and all attempts to intervene through modern forms of gene therapy in order to forestall, say, the development of devastating, inherited conditions or diseases. There is a huge difference between preventing an undeniable harm—say a type of inherited condition that dooms a child to a short and painful life—and striving to create a blemishless perfect human specimen. How one differentiates the one from the other is part of the burden of argument. One example of justifiable intervention would be a method of gene

therapy that spares children "the devastating effects of a rare but deadly inherited disease," described by Denise Grady: "In the condition, Crigler-Najjar syndrome, a substance called bilirubin, a waste product from the destruction of worn-out red blood cells, builds up in the body. . . . Bilirubin accumulates, causing jaundice, a yellowing of the skin and the whites of the eyes. More important, bilirubin is toxic to the nervous system, and the children live in constant danger of brain damage. The only way they can survive is to spend ten to twelve hours a day under special lights that break down the bilirubin. But as they reach their teens, the light therapy becomes less effective. Unless they can get a liver transplant, they may suffer brain damage or die" (Denise Grady, "At Gene Therapy's Frontier: The Amish Build a Clinic," Science Times, *New York Times*, June 29, 1999). Because previous attempts at gene therapy have all fallen far short of expectations, none of this may work. But it would spare a small number of children tremendous suffering, and this sort of intervention is entirely defensive—it involves no eugenics ideology of any kind.

12. This statement from the early HGP leader Walter Gilbert, "On the HGP as the Grail of Human Genetics," is cited in *Forbidden Knowledge: From Prometheus to Pornography*, by Roger Shattuck (New York: Harcourt, Brace 1996), and also reported in Roger Lewin, "Proposal to Sequence the Human Genome Stirs Debate," *Science* 232, no. 4758 (June 27, 1986): 1598–1600. Gilbert's "grail" language has been reported and repeated in many commentaries and books, e.g., Susan Aldridge, *The Thread of Life: The Story of Genes and Genetic Engineering* (Cambridge: Cambridge University Press, 1996; Daniel J., Kevles and Leroy Hood, eds., *The Code of Codes: Scientific and Social Issues in the Human Genome Project* (Cambridge, Mass.: Harvard University Press, 1992); and Roger Lewin, "Proposal to Sequence the Human Genome Stirs Debate," *Science* 232, no. 4758 (1986): 1598–1600, at 1598.

13. As reprinted in "Superior People," editorial, *Commonweal*, March 26, 1999, 5–6.

14. Ibid., 5.

15. Doris T. Zallen, "We Need a Moratorium on 'Genetic Enhancement,'" *Chronicle of Higher Education*, March 27, 1998, A64.

16. James Le Fanu, "Geneticists Are Not Gods," *The Tablet*, December 12, 1998, 1645–46.

17. Ibid.

18. This is a bit reminiscent of Julia, young female sexual revolutionary, in Orwell's *1984* (1949). She is, of course, defeated, and her lover Winston Smith comes to love Big Brother.

19. Jean Bethke Elshtain, "Ewegenics" ("Hard Questions" column), *New Republic*, March 31, 1997, 25. *Note from the author of this chapter and the*

original New Republic *column author (JBE)*: Please note here and further on that I do not want in any way to diminish the difficulties involved in parenting a child with disabilities. As the mother of an adult daughter with mental retardation, I understand this very well. Instead, I am trying to capture the present temperament that dictates that such births are calamitous and ought never to occur.

20. Ibid.

21. Hans S. Reinders, *The Future of the Disabled in Liberal Society: An Ethical Analysis* (Notre Dame, Ind.: University of Notre Dame Press, 2000).

22. Hannemann is quoted in "Search and Destroy Missions," by Kevin Clarke, *U.S. Catholic*, January 2000; http://www.uscatholic.org/2000.01/seades.htm.

23. Stephen S. Hall, "The Recycled Generation," *New York Times Magazine*, January 30, 2000, 30–35, 46, 74–79.

24. Ibid., 32.

25. Ibid.

26. John R. G. Turner, "Review," *Times Literary Supplement*, August 8, 1996, 3. The four books reviewed by Turner were Enzo Russo and David Cove, *Genetic Engineering: Dreams and Nightmares* (Basingstoke: W. H. Freeman, 1995); Philip Kitcher, *The Lives to Come: The Genetic Revolution and Human Possibilities* (London: Penguin Press, 1996); Steve Jones, *In the Blood: God, Genes and Destiny* (London: HarperCollins, 1996); and Susan Aldridge: *The Thread of Life: The Story of Genes and Genetic Engineering* (Cambridge: Cambridge University Press, 1996).

27. Turner, "Review." But who defines excess? This is a squishy soft criterion that now comes into play at present for such "abnormalities" as cleft palate.

28. Sinsheimer is quoted by Roger Shattuck (among others), who picked up on this prediction as the HGP got under way in the 1990s. The original source is Robert L. Sinsheimer, "The Prospect of Designed Genetic Change," *Engineering and Science* 32 [1969]: 8–13. The quotation here is in an overview of "forbidden knowledge," the subject and title of Shattuck's 1996 book; Shattuck, *Forbidden Knowledge: From Prometheus to Pornography* (New York: Harcourt, Brace 1996), 193–94. Sinsheimer's same words on raising genetic fitness through the new eugenics have been examined again recently, and with a cool critical eye, by Michael Sandel, "The Case against Perfection," *Atlantic Monthly*, April 2004, 50–61.

29. The literature of reportage, enthusiasm, concern, etc., is nearly out of control. A few magazine and newspaper pieces worth reading include Jim Yardley, "Investigators Say Embryologist Knew He Erred in Egg Mix-Up," *New*

York Times, April 17, 1999; Martin Lupton, "Test-Tube Questions," *The Tablet*, February 20, 1999, 259–60; David L. Marcus, "Mothers with Another's Eggs," *U.S. News & World Report*, April 13, 1999, 42–44; Nicholas Wade, "Panel Told of Vast Benefits of Embryo Cells," *New York Times*, December 3, 1998; Anne Taylor Fleming, "Why I Can't Use Someone Else's Eggs," *Newsweek*, April 12, 1999, 12; Nicholas Wade, "Gene Study Bolsters Hope for Treating Diseases of Aging," *New York Times*, March 5, 1999; Lisa Belkin, "Splice Einstein and Sammy Glick: Add a Little Magellan," *New York Times Magazine*, August 23, 1998, 26–31, 56–61 (a chilling piece that shows the many ways in which geno-enthusiasm and commodification fuse); and Stephanie Armour, "Could Your Genes Hold You Back?" *USA Today*, May 5, 1999. An example of how the bizarre becomes commonplace is Gina Kolata, "Scientists Place Jellyfish Genes into Monkeys," *New York Times*, December 23, 1999. We have normalized the preposterous and do not even ask, Why on earth would anyone do that—put jellyfish genes into monkeys?

30. This foreboding also comes through in Bryan Appleyard, *Brave New Worlds. Staying Human in the Genetic Future* (New York: Viking Press, 1998).

31. Gina Kolata, "On Cloning Humans, 'Never' Turns Swiftly into 'Why Not?'" *New York Times*, December 3, 1997.

32. National Bioethics Advisory Commission, *Cloning Human Beings: Report and Records of the National Bioethics Advisory Commission* (Rockville, Md.: National Bioethics Advisory Commission, 1997). Laurence Tribe's remarks, originally expressed in a *New York Times* op-ed, are available intact in "Slouching Towards Cloning," by Kay S. Hymowitz, *City Journal*, Winter 1998, along with review of the debate over restricting cloning. See also reports of Tribe's remarks in Kolata, "On Cloning Humans," and other comments on cloning by Tribe in the *New York Times* in "Second Thoughts on Cloning," op-ed, December 5, 1997, and "For Some Human Cloning Might Offer Hope: No Blanket Welcome," *New York Times*, January 14, 1998. Of course, at this writing (2005), things have moved beyond earlier limits, with embryo-destroying "therapeutic cloning" in stem cell research up for grabs in the United States in state legislatures and the U.S. Congress.

33. We cannot here deal with the commercialization of genetics, but it must be noted that the huge profits to be made drive much of the scientific and technological work, alas. E.g., see Belkin, "Splice Einstein and Sammy Glick."

34. Quoted by Kolata, "On Cloning Humans."

35. Think, by the way, of what this would have done to Martin Luther King's protest: simply stopped it dead in its tracks. For the law of the Jim Crow South was the law of segregation. And no ethical argument can challenge the law. End of story. A comeback would be that you need to make a legal argument to change the law. But King's call for legal change was an ethical call. The

reductive argument that law and ethics must never touch is a crude form of legal positivism, or command-obedience legal theory. What is right does not enter into the picture at all.

36. Pontifical Academy for Life, "Reflections on Cloning," *Origins* 28 (May 21, 1998): 14–16; original statement dated June 25, 1997.

37. The popular press has been filled with cloning articles. A few include Sheryl WuDunn, "South Korean Scientists Say They Cloned a Human Cell," *New York Times*, December 17, 1998; Nicholas Wade, "Researchers Join in Effort on Cloning Repair Tissue," *New York Times*, May 5, 1999; and Tim Friend, "Merger Could Clone Bio-Companies' Creativity," *USA Today*, May 5, 1999. Also see Lori B. Andrews, *The Clone Age: Adventures in the New World of Reproductive Technology* (New York: Henry Holt, 1999).

38. But we have a solution to that one, too, don't we? We can be certain that the creatures nobody wants, whose lives are not "worth living," can be easily dispatched to spare their suffering. Physician-assisted suicide, the track down which we are moving, is, of course, part of the general tendencies discussed and criticized here. Although this chapter does not focus specifically on this matter, the following two essays are recommended for the general reader: Paul R. McHugh, "The Kevorkian Epidemic," *American Scholar*, Winter 1997, 15–27; and Leon R. Kass and Nelson Lund, "Courting Death: Assisted Suicide, Doctors, and the Law," *Commentary*, December 1996, 17–29. Recommended, as well, are Joseph Cardinal Bernardin's 1996 "Letter to the Supreme Court," which was appended to a friend-of-the-court brief filed by the Catholic Health Association in a Supreme Court case testing the appeals of two lower court decisions that struck down laws prohibiting assisted suicide in Washington and New York states (*Vacco v. Quill* [New York], U.S. No. 95-1858; *Washington v. Glucksberg* [Washington], U.S. No. 96-110, June 26, 1997), and a brief by the U.S. Catholic Conference, "Assisted Suicide Issue Moves to Supreme Court," *Origins* 26 (December 12, 1996): 421–30.

39. Leon R. Kass, "The Wisdom of Repugnance," *New Republic*, June 2, 1997, 17–26.

40. Ibid., 20.

41. There is a big discussion here yearning to breathe free, of course—namely, the connection between beauty and truth. But it is one we cannot even begin to enter on at this point. The truth is often described as splendid and beautiful—Augustine's language—and God as beautiful in and through God's simplicity. The aesthetic dimension in theology and, most certainly, in ethics is underexplored.

42. Kass, "Wisdom," 20.

43. Ibid., 24.

44. Ibid., 22–24.

45. See Roger Shattuck's wonderful discussion of Faust and Frankenstein in Shattuck, *Forbidden Knowledge*, 79–100.

46. John Paul II, *Original Unity of Man and Woman* (Boston: Daughters of St. Paul, 1981), 23.

47. Karol Wojtyla (Pope John Paul II), *Sign of Contradiction* (New York: Seabury Press, 1979), 24.

48. Ibid., 124.

49. Dietrich Bonhoeffer, *Ethics*, 1st Touchstone ed. (New York: Simon & Schuster, 1995). This discussion appears on 142–85, and all quoted matter is drawn from these pages inclusively.

50. Ibid., 143.

51. Ibid., 144.

52. Ibid., 148.

53. This is an area that deserves longer treatment than can be given here. Fortunately, and at long last, there are texts in English on Nazi euthanasia as part of its general biopolitics. Of especial note is Michael Burleigh, *Death and Deliverance* (Cambridge: Cambridge University Press, 1994). This is a tremendously disquieting book for a contemporary American reader. So much of the language of our own genetic engineering and "assisted suicide" proponents seems to reverberate with echoes of National Socialist propaganda. The Nazis covered the waterfront, so to speak, justifying their programs of systematic selective elimination of the "unfit," of life unworthy of life (congenitally "diseased," handicapped, etc.) on a number of interrelated grounds, including cost-benefit criteria, perfecting the race, and compassion. The Nazis also controlled the media on this issue (it goes without saying), producing short propaganda films and full-length features, lavishly produced and starring German matinee idols, to promote their euthanasian efforts.

54. Bonhoeffer, *Ethics*, 152.

55. Ibid., 153.

56. Martin Luther, "The Gospel for the Early Christmas Service, Luke 2:[15–20]," trans. J. G. Kunstmam, in *Luther's Works*, vol. 52, *Sermons II*, ed. Jaroslav Pelikan (Philadelphia: Fortress Press, 1955), 39–40.

57. Dietrich Bonhoeffer, *Life Together: The Classic Exploration of Christian Community* (Minneapolis: Fortress Press, 1996).

Notes to Chapter 6

1. Plato [ca 380 B.C.], *The Republic*, trans. with notes and an interpretive essay by Allan Bloom (New York: Basic Books, 1968); Thomas More, *Utopia*

(New Haven, Conn.: Yale University Press, 1965 [1516]); Charles Fourier, *Oeuvres Complètes de Charles Fourier* (Paris: Librairie Sociétaire, 1845–46); Francis Bacon, *New Atlantis* (Oxford: Clarendon Press, 1915 [1626]); Aldous Huxley, *Brave New World and Brave New World Revisited* (New York: Harper & Brothers, 1964 [1931]).

2. George Orwell, *1984* (New York: Harcourt, Brace, 1949); Samuel Butler, *Erewhon and Erewhon Revisited* (New York: Random House, 1979 [1927]); Jonathan Swift, *Gulliver's Travels*, ed. with an introduction and notes by Robert DeMaria Jr. (New York: Penguin Books, 2001 [1726]); Anthony Trollope, *The Fixed Period* (Edinburgh: W. Blackwood and Sons, 1882).

3. Lee M. Silver, *Remaking Eden: Cloning and Beyond in a Brave New World* (New York: Avon Books, 1997); Ray Kurzweil, *The Age of Spiritual Machines: When Computers Exceed Human Intelligence* (New York: Viking Press, 1999).

4. R. Ford Denison, E. Toby Kiers, and Stuart A. West, "Darwinian Agriculture: When Can Humans Find Solutions beyond the Reach of Natural Selection?" *Quarterly Review of Biology* 78, no. 2 (June 2003): 146–48; E. O. Wilson, *The Future of Life* (New York: Alfred A. Knopf, 2002); James Watson, with Andrew Berry, *DNA: The Secret of Life* (New York: Alfred A Knopf, 2003); "Humanist Manifesto II," *The Humanist* 33 (September–October 1973): 3–4, 13–14; Francis Fukuyama, *Our Posthuman Future: Consequences of the Biotechnology Revolution* (New York: Farrar, Straus & Giroux, 2002).

5. Sir William Osler, *Man's Redemption of Man: A Lay Sermon*, McEwan Hall, July 2, 1910 (London: Constable, 1913).

6. Lynn White Jr., *Medieval Technology and Social Change* (Oxford: Clarendon Press, 1962), passim.

7. Sheila M. Rothman and David J. Rothman, *The Pursuit of Perfection: The Promise and Perils of Medical Enhancement* (New York: Pantheon Books, 2003).

8. President's Council on Bioethics, *Human Cloning and Human Dignity: The Report of the President's Council on Bioethics* (New York: PublicAffairs, 2002); President's Council on Bioethics, *Beyond Therapy: Biotechnology and the Pursuit of Happiness, A Report of the President's Council on Bioethics* (Chicago: University of Chicago Press, 2003).

9. See Arthur L. Caplan, Tristram Engelhardt Jr., and James J. McCartney, eds., *Concepts of Health and Disease: Interdisciplinary Perspectives* (Reading, Mass.: Addison-Wesley, 1981); and Arthur L. Caplan, James J. McCartney, and Dominic A. Sisti, eds., *Health, Disease, and Illness: Concepts in Medicine*, foreword by Edmund D. Pellegrino (Washington, D.C.: Georgetown University Press, 2004).

10. Leon R. Kass, "Regarding the End of Medicine and the Pursuit of Health," in *Concepts of Health and Disease*, ed. Caplan, Engelhardt, and McCartney, 3–30, at 29.

11. Owsei Temkin and C. Lillian Temkin, eds., *Ancient Medicine: Selected Papers of Ludwig Edelstein* (Baltimore: Johns Hopkins University Press, 1967).

12. Paul Carrick, *Medical Ethics in the Ancient World* (Washington, D. C.: Georgetown University Press, 2001).

13. Galen is quoted by Owsei Temkin, "The Scientific Approach to Disease: Specific Entity and Individual Sickness," in *Concepts of Health and Disease*, ed. Caplan, Engelhardt, and McCartney, 247–63, at 254.

14. Christopher Boorse, "On the Distinction between Disease and Illness," *Philosophy and Public Affairs* 1 (1975): 49–68; Henrik R. Wulff, Stig Andur Pendersen, and Raben Rosenberg, *Philosophy of Medicine: An Introduction* (London: Blackwell Scientific Publications, 1990); Raanan Gillon, *Philosophical Medical Ethics* (New York: John Wiley, 1986); Lawrie Reznek, *The Nature of Disease* (New York: Routledge & Kegan Paul, 1987); Knud Faber, *Nosography: The Evolution of Clinical Medicine in Modern Times* (New York: Paul Hoeber, 1930 [1923]).

15. In *Concepts of Health and Disease*, ed. Caplan, Engelhardt, and McCartney, the examples of different definitions cited here are H. Tristram Engelhardt Jr., "The Concepts of Health and Disease," 31–45; Talcott Parsons, "Definitions of Health and Disease in the Light of American Values and Social Structure," 57–81; Lester S. King, "What Is Disease?"107–18; and Horacio Fabrega Jr., "The Scientific Usefulness of the Idea of Illness," 131–42.

16. The World Health Organization's definition of "health" is in "Preamble to the Constitution of the World Health Organization as Adopted by the International Health Conference, New York, 19–22 June, 1946, Signed on 22 July 1946 by the Representatives of 61 States," *Official Records of the World Health Organization*, no. 2, 100; entered into force April 7, 1948. "Constitution of the World Health Organization," in *Concepts of Health and Disease*, ed. Caplan, Engelhardt, and McCartney, 83–84.

17. Fredrick C. Redlich, "The Concept of Health in Psychiatry," in *Concepts of Health and Disease*, ed. Caplan, Engelhardt, and McCartney, 373–90; Peter Sedgwick, "Illness Mental and Otherwise," in *Concepts of Health and Disease*, ed. Caplan, Engelhardt, and McCartney, 119–29.

18. Bjorn Hofmann, "On the Triad Disease, Illness and Sickness," *Journal of Medicine & Philosophy* 27, no. 6 (December 2002): 651–73.

19. Boorse, "On the Distinction between Disease and Illness."

20. John Worrall and J. Worrall, "Defining Disease: Much Ado about Nothing?" in *Analecta Husserliana LXXII*, ed. Anna-Teresa Tymieniecka and Evandro Agazzi (Dordrecht: Kluwer, 2001).

21. Temkin, "Scientific Approach to Disease," 254.

22. Gregg Easterbrook, *The Progress Paradox: How Life Gets Better While People Feel Worse* (New York: Random House, 2003).

23. Edmond A. Murphy, *The Logic of Medicine* (Baltimore: John Hopkins University Press, 1997).

24. Aristotle, *Nichomachean Ethics: The Basic Works of Aristotle*, ed. with an introduction by Richard McKeon (New York: Random House, 1968), 1094A.

25. Hippocrates, *Hippocrates: The Art*, with English translation by W. H. S. Jones (Cambridge, Mass.: Harvard University Press, 1923).

26. Hippocrates, *Airs, Waters, and Places*, in *Galen's Commentary on the Hippocratic Treatise* (Jerusalem: Israel Academy of Sciences and Humanities, 1981).

27. Edmund D. Pellegrino, "The Goals and Ends of Medicine: How Are They to Be Defined?" in *The Goals of Medicine: The Forgotten Issue in Health Care Reform*, ed. Mark J. Hanson and Daniel Callahan (Washington, D.C.: Georgetown University Press, 1999), 55–68.

28. Ibid.

29. Ibid.

30. E.g., see Erik Parens, "Is Better Always Good?" and Eric Jeungst, "What Does Enhancement Mean?" both in *Enhancing Human Traits: Ethical and Social Implications*, ed. Erik Parens (Washington, D.C.: Georgetown University Press, 1998), 1–28, 29–69, respectively.

31. President's Council on Bioethics, *Being Human: Readings from the President's Council on Bioethics* (Washington D.C.: U.S. Government Printing Office, 2003).

32. Jeungst, "What Does Enhancement Mean?," 29.

33. Dan W. Brock, "Enhancement of Human Function: Some Distinctions for Policymakers," in *Enhancing Human Traits: Ethical and Social Implications*, ed. Parens, 48–69, at 49.

34. Norman Daniels, "Growth Hormone Therapy for Short Stature: Can We Support the Treatment/Enhancement Distinction?" *Growth Genetics and Hormones*, Suppl. 1, no. 8 (1992): 46–48.

35. H. W. Fowler, *A Dictionary of English Usage*, 2nd ed. rev. by Sir Ernest Gowers (New York: Oxford University Press, 1965), s.v. "enhance."

36. Brian Vastag, "Poised to Challenge Need for Sleep, 'Wakefulness Enhancer' Rouses Concerns," *Journal of the American Medical Assocation* 291, no. 2 (January 14, 2004): 167–70.

37. Michael Sokolove, "In Pursuit of Doped Excellence: Lab Animal," *New York Times Magazine*, January 18, 2004, 28.

38. Peter D. Kramer, *Listening to Prozac* (New York: Viking Press, 1993).

39. E.g., see Daniel Callahan, "Visions of Eternity," *First Things* 133 (May 2003): 28–35; and Leon R. Kass, "Ageless Bodies, Happy Souls: Biotechnology and the Pursuit of Perfection," *The New Atlantis* 1 (2003): 19–20. See also Gary Stix, "Ultimate Self-Improvement," *Scientific American* 289, Special Issue, no. 3 (September 2003): 44.

40. Callahan, "Visions of Eternity."

41. See especially Leon R. Kass, "The Wisdom of Repugnance," *New Republic*, June 2, 1997, 17–26; Kass, "Ageless Bodies"; and Leon R. Kass and James Q. Wilson, *The Ethics of Human Cloning* (Washington, D.C.: AEI Press, 1998).

42. Hans J. Achterhuis, "The Courage to Be a Cyborg," in *Research in Philosophy and Technology*, vol. 17, *Technology, Ethics, and Culture*, ed. Carl Mitcham (Greenwich, Conn.: Jai Press, 1998), 9–24.

43. C. S. Lewis, *The Abolition of Man* (New York: Collier Books, 1962 [1947]), 69.

44. See Hans Jonas, *Imperative of Responsibility: In Search of an Ethics of the Technological Age* (Chicago: University of Chicago Press, 1984); and Jacques Ellul, *The Betrayal of the West* (New York: Seabury Press, 1978).

45. Gordon Ethelbert Ward Wolstenholme, ed., *Man and His Future: A CIBA Foundation Volume* (Boston: Little, Brown, 1963).

46. Jonas, *Imperative of Responsibility*, 52.

47. Adam Keiper, "The Nanotechnology Revolution," *The New Atlantis*, no. 2 (Summer 2003): 17–34, at 20.

48. Eric K. Drexler, *Engines of Creation: The Coming Era of Nanotechnology* (Garden City, N. Y.: Anchor Books/Doubleday, 1990 [1986, 1987]).

49. Dan Ferber, "Synthetic Biology: Microbes Made to Order," *Science* 303, no. 5635 (January 9, 2004): 158–61.

50. Jürgen Habermas, *The Future of Human Nature*, trans. Max Pensky and William Rehq (Cambridge: Polity, 2003).

51. Robert A. Freitas Jr., *Nanomedicine, Volume 1: Basic Capabilities* (Georgetown, Tex.: Landesbioscience, 1999).

52. Arthur L. Caplan, "Is Better Best?" *Scientific American* 283, Special Issue, no. 3 (September 2003): 104–5.

53. Ibid.

54. President's Council on Bioethics, *Being Human*.

55. See United Nations, International Bill of Rights, 1966, Incorporating the Universal Declaration of Human Rights, 1948, with the International Covenant on Economic, Social, and Cultural Rights and the International Covenant on Civil and Political Rights. Entered into force January 3, 1976. Background reported in Office of the UN High Commissioner for Human

Rights, Fact Sheet 2, June 1996; http://www.unhchr.ch/the menu6/2/Ps2.htm.

56. Two recent attacks on the dignity concept have come from Ruth Macklin, "Dignity Is a Useless Concept" (editorial), *British Medical Journal* 327, no. 7429 (December 20, 2003): 1419–20; and Matti Hayry, "Another Look at Dignity," *Cambridge Quarterly of Healthcare Ethics* 13, no. 1 (Winter 2004): 7–14.

57. Edmund D. Pellegrino and David C. Thomasma, *For the Patient's Good: The Restoration of Beneficence in Healthcare* (New York: Oxford University Press, 1988).

58. Benedict M. Ashley, *Theologies of the Body: Humanist and Christian* (Braintree, Mass.: Pope John Center, 1995).

59. F. LeRon Shults, *Reforming Theological Anthropology: After the Philosophical Turn to Relationality* (Grand Rapids: Eerdmans, 2003).

60. Martin Rees, *Our Final Hour* (London: William Heinemann, 2003).

61. Ellul, *Betrayal of the West*, 77.

62. Paul Kurtz, Barry Karr, and Ranjit Sandhu, eds., *Science and Religion: Are They Compatible?* (Amherst, N.Y.: Prometheus Books, 2003).

63. Stephen M. Barr, "The Story of Science," *First Things* 131 (March 2003): 16–25.

64. Robert Burton, *The Anatomy of Melancholy* (New York: Tidor Publishing, 1955 [1621]).

65. "Book of Sirach," in *New American Bible* (New York: Bishops Committee of the Confraternity Christian Doctrine Catholic Publishing Company, 1970).

66. Edmund D. Pellegrino, "Bioethics and the Anthropological Question," Geneva Lecture Series Lecture and Colloquium, University of Iowa, Iowa City, April 8, 2005.

67. Pope John Paul II, *Ex Corde Ecclesiae*, no. 18, Apostolic Constitution Promulgated by His Holiness on August 15, 1990; http://www.ewtn.com/library/papaldoc/jp2unive.htm.

Notes to Chapter 7

1. Wolfhart Pannenberg, *Systematic Theology*, trans. Geoffrey W. Bromiley (Grand Rapids: Eerdmans, 1994), vol. 2, 204–5. The charge of ecological destruction leveled against the biblical Judeo-Christian tradition was launched in Lynn White's essay "The Historical Roots of Our Ecological Crisis," in the March 10, 1967, issue of *Science*. Misunderstandings of the scriptural *dominion* granted to humankind have fueled the debate, as Pannenberg explains in his *Systematic Theology*, showing the secular roots of abuse by technology.

2. Robert Kraut, Michael Patterson, Vicki Lundmark, Sara Kiesler, Tridas Mukopadhyay, and William Scherlis, "Internet Paradox: A Social Technology That Reduces Social Involvement and Psychological Well-Being?" *American Psychologist* 53, no. 9 (September 1998): 1017–31.

3. The writings of the Jesuit paleontologist Pierre Teilhard de Chardin are foundational to "evolutionary" theology, including *Christianity and Evolution* (New York: Harcourt Brace Jovanovich, 1971); *The Divine Milieu: An Essay on the Interior Life* (New York: Harper & Brothers, 1957), and in French, *Le Milieu Divin* (Paris: Seuil, 1960); and *The Phenomenon of Man* (New York: Harper & Brothers, 1955), and in French, *Le Phénomène Humain* (Paris: Seuil, 1959). Beyond those who use a general evolutionary paradigm for creation and "development" are those who bring technology into the process, such as Philip Hefner, *Technology and Human Becoming* (Minneapolis: Fortress Press, 2003 [1971]); and Ronald Cole-Turner, *Beyond Cloning: Religion and the Remaking of Humanity* (Harrisburg: Trinity Press International, 2001).

4. Hefner, *Technology and Human Becoming*, 79.

5. Cole-Turner, *Beyond Cloning*, 149–50.

6. American Society of Aesthetic Plastic Surgery, 2002 Statistics, at http://www.surgery.org/download/quick_facts.pdf; online, see "Breast Implants Patients Surveys" (2003).

7. The American Academy of Facial Plastic and Reconstructive Surgery has been tracking rates of cosmetic surgery and the reasons for them; see http://aafprs.org/media/press-release/011501.html.

8. This ad for Pantene Shampoo, featuring model-turned-actress Kelly LeBrock, became popularized with its slogan passing into current parlance, and it has become a classic for analysis. See http://insider.tv.yahoo.com/celeb/insider2005098t11; and "Classic Commercials of the Eighties," http://www.tripletsandus.com/80s/commercials.htm. For a critical analysis of the ad's exploitation of envy, see Gerald Grow (professor of journalism, Florida A&M University, Tallahassee), "Don't Hate Me Because I Am Beautiful: A Commercial in Context," http://www.longleaf.net/ggrow.

9. The Michelle Pfeiffer cover shot in *Esquire* (December 1996) is discussed by Bob Andelman, "Nothing Except $1,525 Touch Ups," http://www.andelman.com/mrmedia/95/7.4.95.html.

10. For information on this "virtual beauty contest," see the Miss Digital World website, http://www.missdigitalworld.com/MDWContest/showpage/6.

11. Suggestions of the American Society of Aesthetic Plastic Surgery are reported at http://surgery.org/press/news-print.php?iid=217§ion=news-psyc.

12. Lewis Mumford, *The Conduct of Life* (New York: Harvest/Harcourt Brace Jovanovich, 1960 [1951]), 124.

13. The history of eugenics in America and the consequences of bad theology not only not opposing but actually promoting the atrocities are well documented in several recent works, including Edwin Black, *War against the Weak: Eugenics and America's Crusade to Create a Super Race* (New York: Four Walls Eight Windows, 2003); Christine Rosen, *Preaching Eugenics: Religious Leaders and the American Eugenics Movement* (New York: Oxford University Press, 2004); Elof Axel Carlson, *The Unfit: A History of a Bad Idea* (Cold Spring Harbor, N.Y.: Cold Spring Harbor Laboratory Press, 2001); and Stefan Kuhl, *The Nazi Connection: Eugenics, American Racism, and German National Socialism* (New York: Oxford University Press, 2002).

14. Egbert Schuurman, *Perspectives on Technology and Culture* (Sioux Center, Iowa: Dordt College Press, 1995), 145.

15. Eric K. Drexler, *Engines of Creation: The Coming Era of Nanotechnology* (Garden City, N.Y.: Anchor Books/Doubleday, 1990 [1986, 1987]).

16. That theistic worldview holds, too, for Islam; but it remains for others to extend that beyond the scope of this book. Shared principles include Creation, sovereignty of God, special human status, and a stewardship relationship with God.

17. In his *Being As Communion: Studies in Personhood and the Church* (Crestwood, N .Y.: St. Vladimir's Seminary Press, 1985), John D. Zizioulas's main thesis is based in trinitarian theology; the quotation here is on p. 18. Even God exists in communion, and as bearers of God's image we too are created for an existence in communion with Him and with other human beings.

18. Mikael Stenmark, *Scientism: Science, Ethics and Religion*, Ashgate Science and Religion Series (Burlington, Vt.: Ashgate, 2001), 1–17.

19. Stephen V. Monsma, *Responsible Technology: A Christian Perspective* (Grand Rapids: Eerdmans, 1986), 49–50.

20. Schuurman, *Perspectives*, 139–40.

21. Dietrich Bonhoeffer, *Ethics* (Minneapolis: Fortress Press, 2005), 91, 178–79.

22. A thorough and excellent discussion of the practical theological implications for the church and each of the members of Christ's Body of the Ascension may be found in *Jesus Ascended: The Meaning of Christ's Continuing Incarnation*, by Gerrit Scott Dawson (Phillipsburg, N.J.: P & R Publishing, 2004).

23. For a helpful discussion of the teleological and eschatological implications of Christology as they relate to technology, see Brent Waters, *From Human to Posthuman: Christian Theology and Technology in a Postmodern World* (Burlington, Vt.: Ashgate, 2006), esp. 106–22.

24. Stephen Williams, "The Theological Task and Theological Method," in *Evangelical Futures: A Conversation on Theological Method*, Papers and Responses from First Annual Theology Conference at Regent College, Vancouver, October 1999, ed. John G. Stackhouse Jr. (Downers Grove, Ill.: InterVarsity Press, 2000 [1999]).

25. The Barna Group, Ventura, Calif., offers many surveys at its website (http://www.barna.org), which present the beliefs of Americans and pastors. E.g., "Survey Reveals the Books and Authors That Have Influenced Pastors," published May 30, 2005, reports a survey of 614 Protestant pastors; it is particularly startling in that only 9 percent of the pastors polled reported being influenced by books on theology, and only 4 percent indicated being influenced by books concerning cultural trends and conditions. Statistics like this are a damning indictment of how unprepared the church is presently for dealing with the challenges it faces.

26. Freeman J. Dyson, *Imagined Worlds* (Cambridge, Mass.: Harvard University Press, 1997), 84.

27. Henri J. M. Nouwen, *In the Name of Jesus: Reflections on Christian Leadership* (New York: Crossroad Publishing, 1989), 20–21.

28. Mumford, *The Conduct of Life*, 232.

Authors and Collaborators

Project Director and Coauthor

C. Christopher Hook, M.D., is consultant in the Division of Hematology and Internal Medicine and the Special Coagulation Laboratory, chair of the Non-Malignant Hematology Group and Clinic, and assistant professor of medicine in the Mayo Medical School, Rochester, Minnesota. He is the former director of ethics education for the Mayo Graduate School of Medicine. He also founded the Mayo Clinical Ethics Council and the Ethics Consultation Service at the Mayo Clinic. He is a fellow of the Institute on Biotechnology and the Human Future of the Chicago-Kent College of Law, Illinois Institute of Technology, and is a member of the U.S. Department of Health and Human Services Secretary's Advisory Committee on Genetics, Health, and Society in Washington. *Dr. Hook's participation and any of his views that might be contained herein are solely his own and do not necessarily express the views of the Mayo Clinic and Foundation or the United States government.*

Authors

Jean Bethke Elshtain, Ph.D., is the Laura Spelman Rockefeller Professor of Social and Political Ethics at the University of Chicago. She is a fellow of the American Academy of Arts and Sciences and the Council on Civil Society. She has served on the Board of Trustees of the Institute for Advanced Study, Princeton, New Jersey, and is currently on the board of the National Humanities Center, Research Triangle Park, North Carolina, and the Board of Directors of the National Endowment for Democracy. She has been a Phi Beta Kappa lecturer and the recipient of nine honorary degrees, and she received the 2002 Frank J. Goodnow Award, the American Political Science Association's highest award for distinguished service to the profession. In 2003 she became the second holder of the Maguire Chair in American History and Ethics at the Library of Congress. She has written sixteen books, including *Public Man, Private Woman: Women in Social and Political Thought* (Princeton University Press, 1993) and *Augustine and the Limits of Politics* (University of Notre Dame Press, 1998). She has written hundreds of articles and essays for popular, public policy, and political science journals. In addition, she gave the 2005–6 Gifford Lectures in Edinburgh.

John F. Kilner, Ph.D., is the Franklin and Dorothy Forman Chair of Ethics and Theology, professor of bioethics and contemporary culture, and director of bioethics programs at Trinity International University in Deerfield, Illinois, and senior scholar of the Center for Bioethics and Human Dignity. He is the author of two books, including *Life on the Line: Ethics, Aging, Ending Patients' Lives, and Allocating Vital Resources* (Eerdmans, 1992), and coauthor or editor of many other volumes.

C. Ben Mitchell, Ph.D., is director of the Center for Bioethics and Human Dignity and associate professor of bioethics and contemporary culture at Trinity International University/Trinity Graduate School. He also edits the journal *Ethics & Medicine: An International Journal of Bioethics*. He is a fellow of the Institute on Biotechnology and the Human Future at the Chicago-Kent College of Law, Illinois Institute of Technology, and a member of the Templeton Oxford Summer Symposium 2003–5. He is author of numerous journal articles, and is the coauthor or coeditor of several volumes, including *Aging, Death and the Quest for Immortality* (Eerdmans, 2004).

Edmund D. Pellegrino, M.D., is the professor emeritus of medicine and medical ethics at the Center for Clinical Medical Ethics at Georgetown University Medical Center, Washington. He was the John Carroll

Professor of Medicine and Medical Ethics and former director of the Kennedy Institute of Ethics, the Center for the Advanced Study of Ethics at Georgetown University, and the Center for Clinical Bioethics. He has authored or coauthored 24 books and more than 550 published articles; was the founding editor of the *Journal of Medicine and Philosophy*; a master of the American College of Physicians; fellow of the American Association for the Advancement of Science; member of the Institute of Medicine of the National Academy of Sciences; recipient of a number of honorary doctorates; and a recipient of the Benjamin Rush Award from the American Medical Association and the Abraham Flexner Award of the Association of American Medical Colleges. In 2004, he was named to the International Bioethics Committee of the United Nations Education, Scientific and Cultural Organization (UNESCO), which is the only advisory body within the United Nations system to engage in reflection on the ethical implications of advances in life sciences. In the fall of 2005, he became chairman of the President's Council of Bioethics.

Scott B. Rae, Ph.D., is currently professor of biblical studies–Christian ethics at Talbot School of Theology, Biola University, La Mirada, California. He is the author many books, including *The Ethics of Commercial Surrogate Motherhood* (Praeger, 1993); *Moral Choices: An Introduction to Ethics* (Zondervan, 2000) and *Brave New Families: Biblical Ethics and Reproductive Technologies* (Baker, 1996). He is also the author of numerous journal articles in bioethics.

Collaborators

Graham Cole, Ph.D., is professor of biblical and systematic theology at Trinity Evangelical Divinity School. He was for ten years principal of Ridley College, University of Melbourne, where he lectured in philosophy, systematic theology, ethics, and apologetics. From 1980 to 1992, he taught at Moore College, Sydney. Two sabbatical years were spent in Cambridge, where he was the Kingdom Hill Fellow at Oak Hill College in London in 1998. He has contributed numerous monographs and articles to books and periodicals. His current research is on the nature of religious experience, eighteenth-century moral philosophy and theology, and the doctrine of God.

Paige Comstock Cunningham, J.D., M.A., is an attorney and educator. Over the past twenty years, she has worked both in private practice and at a public interest law and education organization. She is a member of the board of directors of Americans United for Life, Chicago, and was chair for five years. She is a senior fellow of the Center for Bioethics and Human Dignity and serves on the board of trustees of Taylor University, Upland, Indiana, and the National Advisory Council of the Wheaton College Center for Applied Christian Ethics, Wheaton, Illinois.

Gilbert C. Meilaender, Ph.D., is the Richard and Phyllis Duesenberg Professor of Christian Ethics at Valparaiso University, Valparaiso, Indiana. He is an associate editor for the *Journal of Religious Ethics.* He takes a special interest in bioethics and is a fellow of the Hastings Center, Garrison, New York. His books include *Body, Soul, and Bioethics* (University of Notre Dame Press, 1996) and *Bioethics: A Primer for Christians* (Eerdmans, 1996). He is also a member of the President's Council on Bioethics.

Stephen Williams, Ph.D., is professor of systematic theology at Union Theological College in Belfast. He studied modern history at Oxford University and theology at Cambridge University before completing his Ph.D. in theology at Yale University. Over the years, he has taught courses in various parts of Central and Eastern Europe. He has a long-standing interest in twentieth-century theology in general and Dietrich Bonhoeffer in particular.

Index